THE LABORATORY QUALITY ASSURANCE SYSTEM

THE LABORATORY QUALITY ASSURANCE SYSTEM

A Manual of Quality Procedures
with Related Forms

THOMAS A. RATLIFF, JR.

VNR VAN NOSTRAND REINHOLD
_____ New York

Copyright © 1990 by Van Nostrand Reinhold
Library of Congress Catalog Card Number 89-16495
ISBN 0-442-23459-7

Printed in the United States of America

Van Nostrand Reinhold
115 Fifth Avenue
New York, New York 10003

Chapman & Hall
2-6 Boundary Row
London SE1 8HN, England

Thomas Nelson Australia
102 Dodds Street
South Melbourne, Victoria 3205, Australia

Nelson Canada
1120 Birchmount Road
Scarborough, Ontario M1K 5G4, Canada

16 15 14 13 12 11 10 9 8 7 6 5 4 3

Library of Congress Cataloging-in-Publication Data

Ratliff, Thomas A.
    The laboratory quality assurance system: a manual of quality procedures with related forms / Thomas A. Ratliff, Jr.
      p.   cm.
    Includes index.
    ISBN 0-442-23459-7
    1. Testing laboratories—Quality control—Handbooks, manuals, etc.
I. Title.
    TA416.R37   1989   89-16495
    602.8′7—dc20   CIP

Contents

Preface

The purpose of this book is to provide the user with the means to create a quality assurance manual that will satisfy the needs of his or her particular laboratory, while meeting the requirements of any regulatory or accrediting body with whom the organization may be associated.

The book seeks to arrive at this goal by, first, discussing in detail the requirements of a laboratory quality system as suggested by ANSI/ASQC National Standard Q2-19XX, "Quality Management and Quality System Elements for Laboratories — Guidelines"; second, describing the mechanics of creating a manual; third, providing an example of a laboratory quality manual, complete with sample forms and instructions for their use; and, finally, providing copies of all forms referenced in Part 3, which can be reproduced, copied, or edited to meet the needs of a particular laboratory.

While many of the examples and references included herein relate to industrial hygiene laboratories, the principles set forth are applicable to any testing or analytical laboratory.

Part 1

LABORATORY QUALITY SYSTEM ELEMENTS

SECTION 1   INTRODUCTION

The laboratory may wish to include an Introduction to its Quality Manual to explain its purpose, and how and to whom it is distributed. The Introduction may also be used to explain the background authority or standard on which the document is based and how the manual may differ from such a reference if there is any significant deviation from the authority or standard.

SECTION 2   TITLE PAGE

The Title Page of the manual should contain the following information:

- The name and address of the issuing organization. If the laboratory is a subordinate part of a larger company or corporation, the parent body should be identified, together with its address.
- The name and title of the responsible Quality Control Coordinator or Manager of the Laboratory.
- The name and title of the Laboratory Director, Chief Executive Officer, President of the Corporation, or other individual bearing ultimate responsibility for the quality of the laboratory output.
- The date of issue.
- If distribution of the quality manual is controlled, the copy number of the manual should be indicated on the title page.

SECTION 3   LETTER OF PROMULGATION

If the laboratory desires to issue the quality manual under the cover of a letter of promulgation, this letter should be written and signed by the same person who signed off on the Title Page, i.e., the Laboratory Director, Chief Executive Officer, or Corporate President.

The Letter of Promulgation serves to demonstrate and emphasize management's interest in, and support for, the laboratory's quality program. It should also stress the importance of individual responsibility for conduct which enhances and does not endanger the quality of laboratory performance. Lastly, it should emphasize the fact that the policies and procedures published in the manual are binding on each individual and are the authority as well as the requirement for the conduct of the laboratory's work.

SECTION 4   QUALITY GOALS AND OBJECTIVES

It is the responsibility of the laboratory's management to identify and state, in writing, what the quality goals of the laboratory are to be.

The primary objective of a laboratory's quality system is to improve the precision and accuracy of the laboratory's "product." Here, the laboratory's product can be defined as "the report issued as the result of analytical or testing activity conducted on a sample or samples received from some source." Management, administrative, statistical, investigative, preventive, and corrective techniques are among those which may be employed to maximize the quality of the data.

Secondary objectives which may be established in order to reach primary goals might be:

- To establish the level of the laboratory's routine performance.
- To make any changes in the routine methodology found necessary in order to make it compatible with performance needs.
- To participate in proficiency analytical testing or quality evaluation programs with peer laboratories in order to achieve and maintain consistent, uniform levels of quality.
- To ensure that all personnel are trained to a level of

familiarity with quality technology appropriate to the individual's degree of responsibility.

- To improve and validate laboratory methodologies by participation in method validation collaborative tests.

Quality objectives should be quantified insofar as possible by establishing target dates for completion or by raising a numerical value to a higher level established as a goal. Quality objectives should be attainable. If they are not, they lose effect as a management tool, since they lead to frustration and a resulting lack of enthusiasm and cooperation among those charged with the responsibility for reaching established goals. Quality objectives should be clearly defined and so stated that all concerned understand management's exact intentions with regards to the goals to be reached. Vaguely defined programs, or those not completely understood, are frequently doomed to failure, especially if they are not vigorously supported by management and backed up by adequate follow-up and supervision.

The examples of objectives given above are, by no means, a complete list of quality objectives that might be selected by an individual laboratory. Quality objective should be established based on the particular laboratory's priorities, field of interest, results of audits, or requirements of regulatory or accrediting bodies. Other subjects appropriate for selection as quality goals might have to do with quality costs, seeking accreditation, dealing with customer complaints, etc.

References

Juran, J. M. 1974. *Quality Control Handbook.* New York: McGraw-Hill Book Company.
July, 1983. *Industrial Hygiene Laboratory Quality Control,* Cincinnati: National Institute for Occupational Safety and Health.

SECTION 5   QUALITY POLICIES

Quality policies are established by management in order to arrive at *quality objectives*, meet regulatory or accreditation requirements or, in the case of larger organizations, to agree with previously established policies mandated by higher authority within the company.

Quality policies may be published by standards-writing bodies. Examples are: ISO/IEC Guide 49—"Guidelines for Development of a Quality Manual for Testing Laboratories"; American National Standard ANSI/ASQC Q2-19XX—"Quality Management and Quality System Elements for Laboratories—Guidelines"; and the American Industrial Hygiene Association—"Quality Assurance Manual for Industrial Hygiene Chemistry."

Quality policies may cover such matters as:

- Quality training.
- Publication, distribution, and retention of current or obsolete technical documents such as methods, specifications, calibration procedures, instrument operating instructions, and so on.
- Assurance of good quality, fresh reagents and chemicals, and appropriate calibrated glassware.
- Participation in interlaboratory quality evaluation programs.
- Determination to reduce the costs of correction and evaluation by increasing preventive measures.

These are but a few examples of the kind of policies that may be put forth by management in support of attaining previously established quality objectives. As in the case of published objectives, quality policies should be issued by the highest level of authority available within the laboratory. They should attest to management's concern about and commitment to maintaining a high level of quality in the laboratory's work.

SECTION 6   ORGANIZATION FOR QUALITY

The establishment of a *quality assurance system* in the laboratory, as described here in Part 1, will require the designation of a Quality Control Coordinator within the laboratory to carry out the monitoring, recordkeeping, statistical, calibration, and other functions required in such a program. Other titles, such as "Manager, Quality Assurance," "Director, Quality Assurance," "Quality Supervisor," etc. may be used, but the title "Quality Coordinator" seems particularly appropriate in smaller organizations, such as laboratories are apt to be. Regardless of the title, it is necessary to place the responsibility for carrying out the quality policies prescribed by management on some individual.

This individual may have these duties as his or her sole responsibility in a large organization, and may have a staff, clerical or technical assistance, or, in a smaller organization, may "wear" this position as another "hat."

The Quality Control Coordinator should be placed in the organization at a position where he reports to the highest level at which he can be effective and unbiased in objectively serving the needs of the laboratory. In no case, however, should the quality control coordinating function be subordinate to an individual responsible for direct conduct of testing or analytical work. An example of a typical organization chart for a small analytical laboratory appears as Figure 5-1 in Part 3, XYZ Laboratories, Inc. Quality Assurance Manual. Note that the Quality Coordinator reports to the Laboratory Director. The placement of names in position blocks in the organization charts is optional but often requires unnecessary paperwork, due to frequent personnel changes. In larger organizations, where the Quality Coordinator may have staff support, it is desirable to furnish an additional organization chart for the Quality Assurance Section alone.

The Quality Manual should include a copy of the Quality Control Coordinator's job description, although this may be prepared outside the Quality Control Department. Other position descriptions may be included if required by an accrediting organization. The job description for a Quality Control Coordinator should, as a minimum, include the responsibilities listed in Figure 5-2, in Part 3.

JOB DESCRIPTION

Title: Quality Control Coordinator

1. Basic function:

    The Quality Control Coordinator is responsible for the conduct of the Laboratory Quality Control Program and for taking or recommending measures to ensure the fulfillment of the quality objectives of management and the carrying out of established quality control policies in the most efficient and economical manner commensurate with ensuring the continuing accuracy and precision of data produced.

2. Responsibilities and authority:

    2.1. Develops and carries out quality control programs, including the use of statistical procedures and techniques, which will help the laboratory to meet required or authorized quality standards at minimum cost, and advises and assists management in the installation, staffing, and supervision of such programs.

    2.2. Monitors quality control activities of the laboratory to determine conformance with established policies, regulatory or accreditation requirements, and with good laboratory practice, and makes recommendations for appropriate corrective action and follow-up as may be necessary.

    2.3. Keeps abreast of, and evaluates, new ideas and current developments in the field of quality technology, and recommends courses of action for their adoption or application wherever they fit into the laboratory's area of expertise or policy requirements.

    2.4. Advises the Purchasing Section regarding the quality of purchased supplies, materials, reagents, and chemicals.

    2.5. Supervises the laboratory's interlaboratory proficiency testing program.

    2.6. Monitors the shipping, delivery, packaging, and handling of samples, and makes recommendations for corrective action when conditions are found that lead to damaged, contaminated, or mishandled samples received.

    2.7. Performs related duties as may, from time-to-time, be assigned.

References

Bennett, C. L. 1958. *Defining the Manager's Job.* New York: American Management Association, Inc.

July, 1983. *Industrial Hygiene Laboratory Quality Control.* Cincinnati: The National Institute for Occupational Safety and Health.

January, 1987. *Quality Assurance and Laboratory Operations Manual.* Cincinnati: National Institute for Occupational Safety and Health.

SECTION 7 MANAGEMENT OF THE QUALITY MANUAL

While the laboratory's description of its quality system may be contained in more than one document, the usual format for conveying such information is the *quality manual.* The principal use of the manual is to present, in one document, the laboratory's policies and procedures which relate to the control of the quality of the laboratory's output.

The manual should:

- Include a Table of Contents.
- Assign the responsibility for publishing and distributing the manual and keeping its contents current and up-to-date, and describe the procedures for recommending and making changes.
- Prescribe the format for paragraphing and should maintain the same format uniformly throughout the document.
- Include copies of any reports, forms, tags, or labels which are required to be used by the procedures described in the text of the manual.

If the distribution of the manual is controlled, then each copy should be numbered and a distribution list maintained showing to whom each numbered document has been issued. Unnumbered copies may be distributed on an uncontrolled basis to potential customers, trainees, and the like.

SECTION 8 QUALITY PLANNING

The act of planning is the thinking out, in advance, of the sequence of actions necessary to accomplish a proposed course of action in order to achieve certain objectives. For the quality professional, the task of planning generally presents itself in two phases, discussed below.

The initial phase is encountered when an organization must develop a quality system "from scratch." It is completed when the newly developed system is in place and running. The second phase is a long-term, continuing process, and involves constant appraisal, review, and planning to up-date, improve, or correct deficiencies in the system.

In order for the planner to communicate his plan to the person or persons expected to execute it, he must write it out in the form of procedures, with the necessary criteria, flow-charts, diagrams, tables, forms, etc.

Planning in the field of quality assurance or quality control for the laboratory must fundamentally be geared to the delivery of acceptable quality data at a reasonable quality cost. These objectives are realized only by carefully planning the many individual elements of the quality system which relate properly to each other and are in consonance with the laboratory's overall goals. These elements, taken together, are those discussed in other sections of Part 1. The steps involved in initial quality planning are discussed in detail in Part 2 of this book. The final result of initial quality planning should be a written document which includes the most important information which the planner (normally the Quality Assurance Coordinator) feels should be communicated to users of the document. The resulting overall quality plan then becomes the Quality Manual.

The Quality Assurance Plan, now called the Quality Assurance Manual, has other important functions in addition to the primary purpose already discussed:

- It is the culmination of a planning effort to design into a program or specific project provisions and policies necessary to assure quality data that is accurate, precise, and complete.
- It is an historical record which documents the program or project plans in terms of measurement methods used, calibration standards auditing planned, data validation executed, etc.
- It provides management with a document which can be used as an audit check list to assess whether the quality control and quality assurance procedures called out in the manual are being implemented.
- It may be used as a text book for the training of new employees, or for refresher training.
- It may be used as a sales tool. The existence and use of a Quality Assurance Manual is a powerful sales statement.

The continuing phase of quality planning is kept simmering on the back burner at all times. There should be constant review, appraisal and surveillance of the quality system to seek out and identify departures from procedures specified in the Quality Manual; omissions of expected conduct or neglect to carry out such procedures or the introduction of new, unauthorized procedures into the system. This oversight activity should be carried out in addition to the formal audits periodically conducted by management.

Reference

July, 1983. *Industrial Hygiene Laboratory Quality Control.* Cincinnati: National Institute for Occupational Safety and Health.

SECTION 9 QUALITY IN PROCUREMENT

The laboratory should establish sufficient control over purchased equipment, supplies, chemical reagents, and testing materials to ensure that laboratory operations are not adversely affected by the inadvertent use of substandard equipment or supplies.

PURCHASE ORDERS

A vendor of testing or analytical supplies and materials furnished to laboratories is regarded as a resource to, and an extension of, the laboratory organization. The standards for quality, therefore, required of vendors are the same as those self-imposed on the laboratory.

The purchase order instructs vendors to mark containers of test or analytical materials and reagents and their packing slips with the following information, as applicable:

Name of material
Vendor's name and address
Vendor's lot number
Quantity
Material specification number and date.

This assures that the material is properly identified and that the supplier is using the latest specifications.

Copies of all purchase orders for testing materials, reagents, and analytical goods should be sent to the Laboratory Quality Control Coordinator, where they are reviewed to ensure that the latest requirements are correctly specified.

Purchase orders, receiving documents, and accompanying certifications are used as part of the receiving control procedure, and show information necessary to identify the material being received.

The Laboratory Stores Clerk or another designated person is responsible for checking lots of material received for the correct quantities, for certification, if required, and to check the packing slip against the purchase order. Lots of testing items, such as gas detector tubes received in large quantities, may be subjected to incoming inspection procedures to determine if they meet dimensional and performance specifications. If a discrepancy is found that could affect the quality of laboratory output, the material may be discarded and reordered and the disposition noted on the log sheet. If the material is accepted, the material is logged in and placed in stores, noting:

Identification of the material
Vendor
Date
P. O. number
Assigned log number.

The container label is stamped with the log number and the shelf-life expiration date, if available. No reagents, chemicals, standard solutions, or other time-sensitive materials should be used after the expiration of the assigned shelf-life date.

When, in the judgement of the Quality Control Coordinator, it is desirable to check the validity of a certification of a purchased material, such a check should be made using the laboratory's own expertise and equipment, or by sending the material to an outside laboratory for a third opinion. Such checks should be made at random intervals, or when circumstances dictate the need for a cross-check. In the event of a rejection, the vendor is notified by the Purchasing Department, the material is discarded, and conditions noted.

Procedures should be established to require the Stores Clerk to survey the inventory once monthly, to identify material approaching a shelf-life expiration date within 30 days, so that fresh replacement stocks may be reordered.

As supplies are used or requisitioned, the amounts are posted against the log number and, thus, a running inventory is kept against which replacement orders can be placed by the Stores Clerk to prevent "stock-outs."

Each receiving report is referenced by log number to the applicable purchase order, certification, or report of analytical or test results, and is retained in the quality assurance file.

Logged disposition notes may be reviewed to establish trends in vendor performance, and to ensure a continuing high quality of materials and supplies accepted.

When the Stores Clerk issues materials and supplies to users, they are checked to be sure that the material is properly identified, shows the log number, and has a current shelf-life expiration date. In the case where more than one container of a material is stocked, the oldest is used first, a first-in-first-out (FIFO) regimen.

When the quality, strength, concentration, or composition of reagents, chemicals, solutions or solvents, etc., are always checked against standards or otherwise as part of the method or procedure, there is no need for any check on these materials prior to placing them in stores, other than to validate the identity, shelf-life, or certification, as covered in the paragraphs above.

On occasion, it may be necessary to audit a vendor's quality program to ensure his ability to produce goods to specification. In these cases, a checklist like the Quality System Survey Evaluation Check List (Figure 8-4, Part 3) may be used to evaluate the effectiveness of the vendor's quality system.

The above outline is furnished as a guide to users of this text. It may be changed, added to, simplified, or embroidered upon, as the user sees fit. However, it will be the basis for the Laboratory Quality Control Manual example given in Part 3.

SECTION 10 SAMPLE HANDLING, IDENTIFICATION, STORAGE, AND SHIPPING

Because some samples or their containers are fragile, or are sensitive to environmental changes when shipped from collection points to the laboratory, or are held in storage, special precautions must be taken for handling, storage, packaging, and shipping, to protect the integrity of samples and to minimize damage loss, deterioration, degradation, or loss of identification of the samples.

Physical damage to the sample shipping container may be the fault of the carrier due to mishandling, or it may be the fault of the sender, due to defective or improperly designed packaging.

Sample integrity refers to the cumulative end result of those factors which detract from the overall validity of a field sample. Such factors are:

- Physical damage, as discussed above.
- Loss of the sample due to leakage, breakage of seals, etc.
- Contamination by foreign materials.
- Improper shipping methods of samples requiring special temperature or atmospheric conditions.
- Lack of maintenance of accurate sample identification.

Some analytical or test methods will include detailed instructions and specifications for handling, storage, packaging, and shipping of samples.

The Quality Control Coordinator should be responsible for monitoring this activity, and be responsible for initiating appropriate corrective action when conditions are found which warrant package redesign, or a change in shipping, handling, or storage conditions.

References

July, 1983. *Industrial Hygiene Laboratory Quality Control.* Cincinnati: National Institute for Occupational Safety and Health.

January, 1987. *Quality Assurance and Laboratory Operations Manual.* Cincinnati: National Institute for Occupational Safety and Health.

SECTION 11 CHAIN-OF-CUSTODY PROCEDURES

Chain-of-custody is a term that refers to the maintenance of an unbroken record of possession of a sample from the time of its collection through some analytical or testing procedure, and possibly up to, and through, a court proceeding.

For some laboratories, especially those dealing with forensic and environmental samples, the establishment of chain-of-custody procedures is of paramount importance, as the results of testing or analysis might eventually be held as evidence in a trial or hearing. Such organizations should design their sample-handling documentation systems so that, during each step of sample collection, delivery, receipt, storage, analysis, disposition, or other handling, some one individual is responsible for the custody and the identification of the sample and its accompanying documentation.

The law which governs the chain-of-custody principles was clearly stated by the court in Gallego v. United States, 276 F.2d 914:

"Before a physical object connected with the commission of a crime may properly be admitted in evidence, there must be a showing that such an object is in substantially the same condition as when a crime was committed. Factors to be considered in making this determination include the nature of the article, the circumstances surrounding the preservation and custody of it, and the likelihood of intermeddlers tampering with it. If, upon the consideration of such factors, the trial judge is satisfied that in reasonable probability, the article has not been changed in important respects, he may permit its introduction in evidence."

The statement of the court above refers to chain-of-custody considerations in criminal cases. The principles are essentially the same in civil cases or in administrative law hearings. A laboratory and its personnel may be involved in any one of the three.

Litigation under the Occupational Safety and Health Act of 1970, for instance, and the various environmental acts, is basically civil in nature, but it does have the characteristics of criminal law, particularly when monetary or jail penalties are involved. The legal principles governing chain-of-custody must be adhered to in all three types of litigation.

In any litigation, adherence to chain-of-custody principles has two main goals: (1) to ensure that the sample which is taken or collected, is the same sample which is analyzed or tested; and (2) to ensure that the sample is not altered, changed, substituted, or tampered with between the collection or acquisition and analysis or testing.

If the party attempting to introduce analytical or test results as evidence is challenged by the opposing party, he or she may have to demonstrate an adequate chain-of-custody. This brings up the important question of what constitutes an adequate chain-of-custody. In the Gallego case cited above, the court stated that all that is necessary is that a "reasonable probability that the article has not been changed in important respects" be established. In United States v. Robinson, 447 F.2d 1215 (1971), the court stated that the "probability of misidentification and adulteration must be eliminated not absolutely, but as a matter of reasonable certainty."

It seems clear that absolute security is not necessary for an acceptable chain-of-custody. We will cite one case in which an adequate chain-of-custody was held to be established, and another in which it was not, in order to present to the reader a frame of reference.

In Ohio v. Conley, 288N. E. 2d, 296 (1971), the defendant contested the admission of certain orange pills as evidence claiming that an adequate chain-of-custody of the pills had not been established by the state. The pills had been confiscated from the defendant by a police officer, transferred to another officer, and finally transferred to the laboratory. It is at this point that the defendant claimed that the chain-of-custody was violated. The officer testified that he gave the container of pills to a man in the laboratory, who then assigned it a number. The next testimony was from the chemist who removed the sample from a file drawer and analyzed it. The individual who accepted the sample at the laboratory was not identified nor did he testify. The exact procedure by which the container found its way into the file cabinet is unknown. The court held that even though direct testimony regarding the period in question was not available, an adequate chain-of-custody was established by inference. The container obtained by the analyst was the same as that supplied by the police, and it contained the same number of orange pills. The inference is strong that it was the same container and pills.

In Erickson v. North Dakota Workmen's Compensation Bureau, 123 N. W. 2nd 292 (1963), a coroner removed a blood sample from a deceased individual, placed it in an unsealed container, and transported it to a hospital, where it was given to an unidentified emergency room attendant with instructions that it be placed in a refrigerator. The record disclosed that some time before noon on the following day, a laboratory supervisor found the tube in a refrigerator and had it analyzed for its alcohol content. No one testified as to the time the sample was placed in the refrigerator. The refrigerator was not secured or in a secure area and was accessible to the entire hospital. The court held that the chain-of-custody was defective because the character of the sample could have changed if not refrigerated promptly and it could have been tampered with while in the uncontrolled refrigerator.

Other cases which discuss chain-of-custody are Perry v. Oklahoma City, 470 P. 2d 974 (1970), which held that an adequate chain-of-custody was established; Todd v. United States, 384 F. Supp. 1284 (1975), and Unigard Insurance Company v. Elmore, 224 S. E. 2d 763 (1976), both held that the chain-of-custody was deficient and the evidence was not admissable.

Reference

July, 1983. *Industrial Hygiene Laboratory Quality Control.* Cincinnati: National Institute for Occupational Safety and Health

SECTION 12 LABORATORY TESTING AND ANALYSIS CONTROL (Intra- and Interlaboratory Testing)

All laboratories must establish some means to ensure that testing and analytical procedures are operating within reasonable control. In order to do this, laboratories engage in intra- and interlaboratory testing programs, make sure that they use rugged, published, approved methods (where available), and that these are employed under controlled conditions. Furthermore, adequate, complete records of testing and analytical results obtained as a result of such testing programs must be documented and retained.

In addition, there are the many statistical techniques and control procedures discussed in quality control text books. However, most of these techniques and procedures are found to be useful in manufacturing operations. Unfortunately, laboratories must contend with a variety of obstacles which are rarely encountered in manufacturing.

For example, $\overline{X}$-R charts used in recording and controlling manufacturing operations are usually based upon a large volume of data generated over a relatively short period of time. However, a laboratory, especially one that performs nonroutine analysis or testing, may take years to develop an adequate data base. In this instance, one of two approaches may be taken: (1) a short-term study is conducted to evaluate the variability of data generated using statistical tests which are more useful on a one-time basis, such as t-tests, F-tests, or ANOVA, or (2) trial statistical control limits are calculated using variability estimates based on prior or published results for similar analytical or test methods or calculations that use the error estimated for each testing or analytical step.

Most manufacturing processes involve infrequent process changes, whereas the analytical chemist or test engineer must frequently deal with samples which differ from specified standard products or, in the case of samples to be analyzed, which have different concentration levels and interferences requiring test or method modification. These method modifications can affect the precision and accuracy of results, and produce what appear to be out-of-control points on standard quality control charts. Careful selection of the variables to be charted, an understanding of the method limitations, comparison of results against previously used, independent methods, and active participation in available proficiency testing programs or an exchange of samples between laboratories become important additions to normally used control charts for the scientist, technician, or test engineer.

In manufacturing processes, calibration is not usually a significant source of error, whereas in the laboratory, calibration errors may be the largest source of error. In some instances, these errors may be hidden and related to: (1) limitations in knowledge and agreement over what constitutes the best calibration standard available, (2) errors (both systematic and random) in primary calibration standards, and (3) errors (both systematic and random) inherent in the preparation of working standards. Silica calibration is an example of the first limitation. Universal agreement on what is the most appropriate calibration to use for "respirable dust" silica determinations is not available. This is made important since silica determinations by all three common methods of analysis, i. e.: colorimetric, infrared, and X-ray diffraction, have been reported to have a particle size dependence. The occurrence of certified, calibration-grade gas cylinders having out-of-specification contents is an example of the second limitation. The Environmental Protection Agency has reported on the existence of such cylinders, which are commercially available, and the National Institute for Science and Technology (NIST) has reported problems with the reliability of the low ppm cylinders initially tested in the NIST Standard Reference Material Program. Calibrations involving gases at normal ambient temperature and pressure, such as vinyl chloride, are an example of the third limitation. Because it is difficult to measure the volume of a gas and prevent its loss during the preparation of secondary and working standards, large inaccuracies can occur. When calibration procedures must deviate from accepted practices as taught in college chemistry courses to rely on gas-versus-liquid-versus-solid measurements, and fail to use a consecutive dilution of standards to the working range, sizeable calibration errors are probable.

The use of NIST standard reference materials, even though expensive, the verification of commercial standards by a comparison with NIST standards, proper identification and cross-referencing of the standards used in calibration, the expiration dating of all standards, participation in proficiency testing or calibration procedures, and adherence to proper, written calibration procedures are especially important when one considers that many calibration errors are hidden and can affect laboratory results over a long time period.

Another obstacle is the limited information which

may be available to the test engineer, technician, or analyst about the nature of the samples presented for analysis or testing. In the case of analytical samples; without information on what concentration levels or interferences to expect, gross analytical and calculation errors, such as a failure to compensate for interferences, errors in dilution, or misplaced decimal points may occur, which go undetected when analytical results on field samples are produced and used. In the case of product samples, submitted for physical testing, a lack of pertinent information about the nature of the sample, its source, its intended use, or (perhaps) its history, may lead to the selection of the wrong test method, resulting in the production of useless test data. Even when errors are suspected, repetition of tests on, or analyses of the sample submitted is not possible. This makes careful checking of the procedure, independent verification of calculations, and the use of testers or analysts who are familiar with the product or process being investigated, all-important.

Perhaps the largest advantage that manufacturing quality control efforts have over laboratory quality control efforts is that management has perceived that improved quality leads to reduced costs and higher profitability. Laboratories, as a rule, have been slow to adopt quality cost reporting as a routine management tool. Conversely, in the manufacturing community, a common practice and a requirement of one of the most widely used quality standards used in the United States, MIL-Q-9858A "Quality Program Requirements," is to report on and relate the cost of the control of quality to the savings resulting from those efforts. This approach increases the probability that costs incurred in the conduct of quality control programs will be accepted by top management, since they are deemed to be a cost-saving measure. Perhaps the largest obstacle to improving laboratory quality is the gap in the communication chain between the producer or generator of the sample and the laboratory. Improvements in measurement reliability, precision, and accuracy go unnoticed in the field. This lack of communication between the laboratory and the supplier of the sample to be tested or analyzed results in field personnel being unaware of the limitation of the data. In order to improve communication, laboratory personnel should report limits of detection, confidence limits, and make qualifying statements, when necessary or appropriate, to make sure that laboratory results are not misinterpreted or misused. Users, on the other hand, should take a skeptical look at laboratory results. Submission of blind, split, spiked, and reference samples should be routine when it is possible to provide such test samples. User requirements that laboratories providing analytical and physical testing services participate in proficiency and laboratory accreditation programs and present information on their quality control procedures will provide some

assurance that minimum performance standards can be met by the laboratory.

As can be seen from the above discussion, participation in interlaboratory testing programs is a vital part of the laboratory quality program. Furthermore, such participation may be a requirement of an accreditation program. While there are more than 150 bodies offering accreditation status for laboratories in the United States, not all have a requirement that the laboratory have a quality program in being. Some that do are the American Industrial Hygiene Association (AIHA), 475 Wolf Ledges Parkway, Akron, OH 44311, The Joint Commission on Accreditation of Health Care Organizations, 875 North Michigan Ave., Chicago, IL 60611, and The American Association for Laboratory Accreditation (A2LA), 656 Quince Orchard Road, Gaithersburg, MD 20878.

While there are numerous proficiency testing programs established in both the public and private sectors, some that may be of interest to readers follow:

(1) The Occupational Safety and Health Administration (OSHA) lead (Pb) standard, 29CFR 1910.1025(j)(2)(iii), requires blood lead analyses to be performed by an approved laboratory participating in the Centers for Disease Control blood lead proficiency testing program. For information concerning this program, as well as other proficiency testing programs in microbiology, immunology, immunohematology, and chemistry, one should contact: Proficiency Testing Branch, Centers for Disease Control, Bldg. 6, Room 315, Atlanta, GA 30333.

(2) The Environmental Protection Agency has programs in microbiology, radiochemistry, water pollution and water supply, and interlaboratory audits for air sources, ambient air analyses, and bulk asbestos identification. Information may be obtained by contacting: U.S. EPA Environmental Monitoring and Support Laboratory, Research Triangle Park, NC 27711.

(3) The National Institute for Occupational Safety and Health (NIOSH), Robert A. Taft Laboratories, 4676 Columbia Parkway, Cincinnati, OH 45226 (Attn: MCRB, DPSE), through its Proficiency Analytical Testing (PAT) Program, provides reference samples to public and private industrial hygiene laboratories, with the American Industrial Hygiene Association handling the arrangements.

Although one objective of PAT is to determine whether participating laboratories can perform selected sample analyses, and obtain results within acceptable limits, the primary objective of the program is to upgrade and improve the analytical performance of participating laboratories.

In order to illustrate how a typical interlaboratory testing program operates, we will discuss the NIOSH PAT Program in detail.

In 1972, The PAT Program was started as a proficiency testing program for laboratories providing analytical services to NIOSH and OSHA to insure the agreement of results from the several laboratories reporting data in the Target Health Hazard Program. Initially, PAT provided reference samples of lead, silica, and asbestos, three of the five substances considered to be major health hazards in the Target Health Hazard Program (THHP), to participating laboratories every two weeks for each analyst in each laboratory doing THHP analyses. The program was almost immediately expanded to allow other government and university laboratories to participate. Within a year, it became evident that guidelines establishing minimum standards for personnel, facilities, equipment, record-keeping, and internal quality control were necessary to improve analytical performance. Validation of previously volunteer-developed criteria by two American Industrial Hygiene Association (AIHA) ad hoc committees and the subsequent formal AIHA Laboratory Accreditation Committee was supported by NIOSH contract. Later in 1972, NIOSH provided the funding for the development of validation criteria by AIHA, and the AIHA Laboratory Accreditation Program became operational in 1974, with NIOSH providing the PAT program, in which participation was required for laboratories seeking accreditation.

Now the AIHA handles the arrangements for the provision of a single sample kit each quarter to each of the participating public and private laboratories. Because the frequency of testing has been reduced from once every two weeks to once every quarter, and from evaluating every analyst performing a particular type of analysis each time to rotating sample kits among all analysts performing similar analyses, the PAT Program is designed to complement, but not replace, the laboratory's internal quality control system.

The NIOSH Division of Physical Sciences and Engineering (DPSE) was responsible for the preparation and submittal of reference samples, which presently include the following materials: lead, silica, asbestos, cadmium, zinc, and one of the following eight organic solvents: toluene, benzene, carbon tetrachloride, chloroform, ethylene dichloride, paradiozane, trichloroethylene, and xylene. Sample generation, data processing, and preliminary data evaluations are performed by a contractor to NIOSH specifications.

The use of no specific method is required; however, procedures will be furnished to participating laboratories upon request.

Samples are submitted at four concentration levels and a blank. These levels are selected as representative of concentrations that would be collected under actual field conditions for normal sampling intervals. They are designed to span the threshold limit values for the particular materials.

Not all laboratories in the program analyze all sample sets, nor are they required to. Laboratories are asked to analyze the samples within 15 working days of receipt and send the results to the contractor's computer center for preliminary evaluation. The Monitoring, Control, and Research Branch of DPSE then evaluates these results in more detail, screens laboratories for those with questionable performance, and notifies each laboratory of its status. Proficiency is determined on the basis of a laboratory's performance compared to the performance of all peer laboratories. A chart for each laboratory for each material is maintained, which shows its performance over time.

Prior to any material being included in the PAT Program, it must have undergone preliminary testing to ensure that uniform samples can be prepared, that satisfactory analytical procedures are available, and that samples have a satisfactory shelf life and ruggedness for the program. While accuracy is not a prerequisite for proficiency, it is a parameter that is not overlooked.

By contract with the National Institute for Science and Technology, multiple sets of PAT samples are periodically analyzed for determination of the precision and accuracy with which sample sets are generated. This identifies analytical problem areas and aids in the evaluation of participating laboratory performance.

As stated above, proficiency test data are evaluated by a comparison of an individual laboratory's results with the results of the entire group performing that analysis. Information on intralaboratory variation is based on computation of filter (but not charcoal tube) residual values for each laboratory. These residual values are estimates of intralaboratory variation expressed as the difference between the observed result and an expected result free of excessive intralaboratory variation. Similarly, laboratory-to-laboratory variation is provided by comparison of a specific laboratory's mean result for the four filters or charcoal tubes with the means of all laboratories.

Charts are maintained for residuals and means for each laboratory for each contaminant. An example for the asbestos count results is shown in Figure 1-12-1. This is a type of control chart, and may be interpreted in much the same manner. Limits are calculated and shown on the chart to provide guidelines by which a laboratory may judge its performance compared to that of other participating laboratories. These limits identify those results which differ significantly from the majority of the data points and, as in the case of control charts, the implication follows that there must be an assignable cause which led to an extreme result, or outlier.

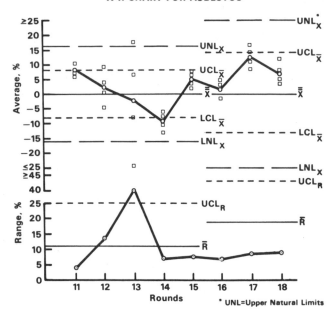

Figure 1-12-1

The PAT Program has, in common with all such interlaboratory test programs, two objectives: The immediate and obvious goal of determining whether the participating laboratories can perform selected sample analyses and obtain results within acceptable limits; and a continuing and underlying objective of the upgrading and improvement of the analytical performance of the laboratories involved.

A special form of interlaboratory testing is collaborative testing. Laboratories participating in collaborative tests use the same measurement method procedure, and every effort is made by the initiating or coordinating laboratory to be sure that all measurement parameters are controlled to the same extent by each laboratory. Collaborative testing is commonly used to measure the total performance of a measurement method, both sample collection and the analysis or testing in terms of bias (accuracy) and precision (reproducibility, repeatability, and replicability). Typically, a collaborative test might consist of the steps listed following.

1. Ruggedness testing. The coordinating laboratory conducts a series of tests to determine the sensitivity or to confirm the insensitivity of the total measurement system, both sample collection and testing or analysis, to variations of certain parameters suspected of affecting the measurement system. The method writeup is reviewed and modified to improve clarity and completeness, particularly with regard to sensitive measurement parameters determined from the ruggedness testing.

2. Collaborative testing. The coordinating laboratory arranges for a minimum of ten laboratories and/or

analysts to participate in a carefully planned test program. Each laboratory provides its own sample collection and analytical or testing equipment. The test program is designed to determine measurement method bias (accuracy) and precision (reproducibility, repeatability, and replicability). Sample collection and analysis or testing is performed on both reference standards and test samples, and comparisons are made to determine the deviation of method results from those made on the reference standard materials.

With regard to intralaboratory testing programs, the purpose of such activity is to identify the sources of measurement method error and to estimate their bias (accuracy) and variability (repeatability and replicability). For manual measurement methods, in the cases where sample collection is followed by laboratory analysis or tests, bias and variability are determined separately for sample collection and analysis and then combined for total method bias and variability. Where continuous data recording is involved, total method bias and variability are determined directly. Some of the potential error sources are the operator, the analyst or test technician, the equipment, the calibration, and the operating conditions. The results may be analyzed

PROBLEMS IN ASSESSING ANALYST PROFICIENCY

Problem	Solutions and decision criteria
Kinds of samples	1. Replicate samples of unknowns or reference standards 2. Consider cost of samples. 3. Samples must be exposed by the analyst to same preparatory steps as are normal unknown samples.
Introducing the sample	1. Samples should have same labels and appearance as unknowns. 2. Because checking periods should not be obvious, supervisors and analyst should overlap the process of logging in samples. 3. Supervisor can place knowns or replicates into the system occasionally. 4. Save an aliquot from one day for analysis by another analyst. This technique can be used to detect bias.
Frequency of checking performance	1. Consider degree of automation. 2. Consider total method precision. 3. Consider analyst's training, attitude, and performance record.

Figure 1-12-2

10

by making comparisons against each other or against reference standards. Operator or analyst proficiency is an additional consideration for intralaboratory testing. While many of the techniques employed in the conduct of an *interlaboratory* testing program are applicable in a modified form to *intralaboratory* testing, there are additional problems related to in-house proficiency testing of operators, testers, or analysts. The major problems associated with designing a program to audit the analyst's or tester's proficiency are concerned with the following:

1. What kinds of samples to use.
2. How to prepare and introduce the samples into the run without the recipient's knowledge.
3. How often to check the analyst's or tester's proficiency.

These problems, and suggested solutions or criteria for decisionmaking, are listed in Figure 1-12-2.

References

Decker, C. E., Murdock, R. W., Arey, F. K. February, 1979. *Final Report—Analysis of Commercial Cylinder Gases of Nitric Oxide and Sulfur Dioxide at Source Concentrations—Results of Audits 1 & 2.* Research Triangle, NC: Research Triangle Institute, (Prepared for the Environmental Protection Agency—EPA Contract No. 68-01-2725)

Decker, C. E. and Encke, R. E. September, 1979. *Final Report—Analysis of Commercial Cylinder Gases of Nitric Oxide, Sulphur Dioxide and Carbon Monoxide at Source Concentrations—Results of Audit 3.* Research Triangle Park, NC: Research Triangle Institute, (Prepared for the Environmental Protection Agency—EPA Contract No. 68-02-3222)

Ho, James Shou-Yien. March, 1979. "Collaborative Study Reference Vinyl Chloride Charcoal Tubes." *American Industrial Hygiene Association Journal,* **40**:200.

16 December, 1963, MIL-Q-9858A *"Quality Program Requirements"* Washington, D.C.: Department of Defense.

July, 1983. *Industrial Hygiene Laboratory Quality Control.* Cincinnati: The National Institute for Occupational Safety and Health.

December, 1976. *Technical Report No. 78, Industrial Hygiene Laboratory Quality Control Manual.* Cincinnati: National Institute for Occupational Safety and Health.

1975. *Quality Assurance Handbook for Air Pollution Measurement Systems.* Research Triangle Park, NC: U. S. Environmental Protection Agency.

SECTION 13 QUALITY DOCUMENTATION AND RECORDS

The laboratory quality assurance system program should include a system for maintaining necessary records and reports and for updating and controlling the issuance of technical documents and operating procedures.

DOCUMENT CONTROL

The important elements of the quality assurance program to which document control should be applied include:

1. Sampling procedures
2. Calibration procedures
3. Analytical and test methods
4. Data collection and reporting procedures
5. Auditing procedures
6. Sample shipping, packaging, receiving, and storage procedures
7. Computation and data validation procedures
8. Quality assurance manuals
9. Quality plans
10. Sampling data sheets
11. Standards

Each laboratory should maintain full control over the distribution and possession of such documents. A file control should be established within the organization showing the following minimum information:

- Document number
- Title
- Source of the document
- Latest issue date
- Change number
- List of addressees

Whenever a change is made, the responsible organization should issue the new, changed document together with the change notice. (See Figure 12-1, Part 4). Whenever practicable, recipients of new or changed documents should acknowledge receipt by signature. Obsolete documents should be removed from points of use and destroyed immediately. Originals of changed documents and their accompanying change notices should be retained in a master file for historical purposes.

Requests for technical document changes, such as changes to methods, sampling data sheets, calibration procedures, and the like can be initiated by anyone within the organization, the request being made in writing on the Technical Data Change Notice (Figure 12-1). It should go through established approval procedures before publication and distribution.

Changes may be promulgated by (1) the issuance of entire new documents, (2) the issuance of replacement pages or, in the case of minor changes, correction of errata, etc., by (3) pen and ink posting on the original document, with this action noted on the change notice. The Quality Control Coordinator should be designated as the responsible individual for ensuring that up-to-date documents are being used and that obsolete documents have been removed from use.

QUALITY RECORDS

Records of other laboratory activities should be maintained in addition to those generated by use of the forms and reports listed above. Such records should be controlled by specifying the individual or organization responsible for their preparation, distribution, and maintenance; the format in which they are published and maintained; the distribution list and the retention period. Such records include:

- Test and analytical results
- Reports on the results of data validation
- Internal and external quality audits
- Instrument and gage record cards
- Quality cost reports
- Laboratory notebooks
- Sample chain-of-custody records

While being stored for specified or required retention periods, documents should be protected from damage, tampering, loss, or degradation due to atmospheric conditions.

LABORATORY NOTEBOOKS

The laboratory notebook is the primary source for documentation of the individual analyst's, test engineer's, or technician's activities. Laboratory notebooks are used for recording all experimental, testing, and analytical notes and data.

The issue of notebooks should be controlled by assignment of an individual serial number to each book. Notebooks are issued to individuals and the serial number noted in an Issue Log. The serial number is entered on the cover of each notebook together with the recipient's name and the date the notebook was issued. Upon completion the notebook is returned for filing and the completion date noted on the cover, Notebooks are hardcovered and bound. Notebooks with removable pages (e.g., loose-leaf notebooks) or loose inserts in the notebook are not considered by many to be acceptable for use in the laboratory. All entries should be made in ink. The pages of the notebook should be numbered and dated and any entries made by an individual other than the person to whom the book was assigned should be noted. The notebooks should be considered to be the property of the laboratory and are retained as part of the laboratory's files.

The notebook should contain all the information gathered by the analyst or test technician pertaining to the sample, including method response (raw data) for each sample. Where appropriate, laboratory number, field number, sequence number, or other identifying numbers should be noted. Analysis or test requested, identification of the method, if known, modifications to the method and the sample originator should also be included. A description of the sample, in as much detail as possible, should be included. Data such as blank values, recovery studies, or duplicate determinations should also be included. Formulas and equations used to calculate results and all calculations should be shown.

The minimum data and description entered in the notebook should be sufficient to enable another test engineer, technician, or analyst to derive the same results as the original worker, with no other source of unpublished information. In addition to these minimum data, any other facts pertinent and appropriate to the sample test or analysis should be entered.

Deletion of errors should be made by drawing a single line through the error. The line drawn should not render the deletion illegible. A notation stating the reason for the deletion should be added and initialed by the person who made the deletion. The recording of data on loose sheets for later entry into the bound notebook is poor procedure and, because of the possibility of transcription errors, should be avoided.

An analyst's or test technician's notebook is always subject to inspection by his colleagues, supervisors, or site visitors, assessors, surveyors, or auditors from outside the laboratory. Therefore, it is imperative that the notebook be maintained in a professional manner and contain all of the pertinent information that may be required by other parties, regardless of the particular importance of that information to the analyst or technician. Furthermore, the notebook must be maintained in such a manner that it can withstand challenges as to the validity, accuracy, or legibility of its contents. Entries should be timely, and not accumulated for more than one day.

References

July, 1983. *Industrial Hygiene Laboratory Quality Control.* Cincinnati, OH: National Institute for Occupational Safety and Health.

1975. *Quality Assurance Handbook for Air Pollution Measurement Systems.* Research Triangle Park, NC: U.S. Environmental Protection Agency.

January 1987. *Quality Assurance and Laboratory Operations Manual.* Cincinnati, OH: National Institute for Occupational Safety and Health.

SECTION 14 CONTROL OF MEASURING AND TEST EQUIPMENT

GENERAL

Calibration procedures require the application of primary or secondary standards. The standards used, whether they are physical or reagent standards, should be certified as being traceable to standards of the National Institute of Science and Technology (NIST) (formerly, the National Bureau of Standards–NBS), or some other recognized fundamental standard. This kind of traceability is necessary even when the standards are generated in the laboratory. Regardless of the type of calibration equipment or material, an effective quality assurance program requires accuracy levels of the standards that are consistent with the test or analytic method.

Calibration procedures apply to all instruments and gages used for analyses and tests, the results of which are recorded for purposes of decision making. The standards used in the calibration of instruments and gages are also included in the calibration system. Instruments not included in the calibration system are those used as indicators only. An example might be a panel voltmeter which indicates when a switch is moved to the "on" position, and whose reading (118 volts, for instance) is not recorded. Indicating instruments should be tagged as such.

CALIBRATION PLAN

A detailed plan should be provided for controlling the accuracy of measuring and test equipment, software and calibration standards used in doing calibration work.

The plan should include:

1. A listing of all required calibration standards with proper nomenclature and identification numbers assigned.
2. The environmental conditions (temperature, relative humidity, barometric pressure, etc.) to be maintained by the calibration activity under which the calibrations standards will be used and the calibrations performed.
3. Established, realistic calibration intervals for measuring and test equipment, and, for each calibration standard, designation of appropriate calibration sources.
4. Written calibration procedures for measuring and test equipment and calibration standards, including document control numbers for reference purposes.
5. A description of the mechanism used to establish traceability of calibration standards to standards available at the National Institute for Science and Technology, or other recognized fundamental standards.
6. A description of the laboratory calibration system, describing how gages and instruments are recalled in a timely manner for scheduled calibration, and including samples of labels, decals, record cards, etc., used in the calibration record system.

CALIBRATION STANDARDS QUALITY

Transfer standards should have four to ten times the accuracy of field and laboratory instruments and gages. For example, if a thermometer used in the laboratory to determine a solution temperature has a specified accuracy of $\pm 2\,°F$, it should be calibrated against a standard thermometer with an accuracy of $\pm 0.2\,°F$. The calibration standards used in the measurement system should, in turn, be calibrated against higher level, primary standards having unquestionable and higher accuracy. These primary standards, in turn, should be certified by NIST or another recognized organization or derived from accepted values of physical or chemical constants.

Calibration gases purchased from commercial vendors normally contain a certificate of analysis. Whenever a certified gas is available from the National Institute for Science and Technology, commercial gas sources should be requested to establish traceability of the certificate of analysis for the certified gas. Inaccurate concentrations in certified gases may result in serious errors in reported measurements of concentrations undergoing analysis or test.

ENVIRONMENTAL CONDITIONS

Measuring and test equipment and calibration standards should be calibrated in an area that provides for control of environmental conditions to the degree necessary to assure the required accuracy. The calibration area should be reasonably free of dust, vapor, vibration, and radio frequency interferences; and it should not be close to equipment that produces noise, vibration, or chemical emissions, or close to areas in which there is chemical testing or the use of microwave or radar transmissions.

The laboratory calibration area should have adequate temperature and humidity controls. A temperature of $68\,°F$ to $73\,°F$ and a relative humidity of 35 to 50 percent normally provide a suitable environment.

A filtered air supply is desirable in the calibration area. Dust particles are more than just a nuisance; they can be abrasive, conductive, and damaging to instruments.

Other environmental conditions which should be considered are:

1. Electric power. Recommended requirements for electrical power for laboratory use should include voltage regulation to within at least 10 percent (preferably 5 percent) of nominal, and minimum line transients, as may be caused by interaction with other users on the main line to the laboratory. Separate input power should be provided, if possible. A suitable grounding system should be established to

assure equal potentials to ground throughout the laboratory.

2. Lighting. Adequate lighting at suggested values of 80 to 100 foot candles at bench levels should be provided. Fluorescent lights should be shielded properly to reduce electrical noise.

CALIBRATION INTERVALS

All newly acquired gages and instruments, as well as those which have been repaired, rebuilt, or reconditioned, should be calibrated prior to being put into use. After initial calibration, all calibration standards and measuring and test equipment should be assigned an established interval for calibration (Figure 13-1, Part 3). In the absence of a published, established calibration interval based on equipment manufacturer's recommendations, authorized government specifications, etc., for a particular item, an initial servicing interval should be assigned by the laboratory or calibration service. The calibration intervals should be specified in terms of time or, in the case of certain types of test and measuring equipment, period of use, or number of times cycled.

The establishment of prescribed intervals should be based on the inherent stability or sensitivity of the equipment, its purpose or use, accuracy, and conditions or severity of use. The intervals may be shortened or lengthened by evaluating the results of the previous and present calibrations and adjusting the schedule to reflect the findings. These evaluations and resulting adjustments must provide positive assurance that changes to calibration intervals will not adversely affect the accuracy of the system.

The laboratory should maintain proper usage data and historical records for all test and measuring equipment, to ascertain whether an adjustment of the calibration is warranted.

Adherence to the calibration frequency schedule is mandatory. Prior to the date when the item is due for scheduled calibration service, it is recalled and removed from service. The recall system may be a simple tickler file, with Instrument/Gage Calibration Records (Figure 13-2, Part 3) being filed by month, in laboratories having a small gage and instrument inventory. In this case, the cards for items due for calibration in a given month are pulled on the first of that month and the gages or instruments are recalled for calibration. In the case of organizations with large inventories, it is common to computerize the recall system and publish computer printouts, which are distributed to all affected areas, to initiate the scheduled recalls.

On occasion, it may be necessary to calibrate between normal, scheduled calibration due dates, if there is evidence of damage or suspected or apparent inaccuracy in the equipment.

CALIBRATION PROCEDURES

Written step-by-step procedures for the calibration of measuring and test equipment and calibration standards should be used by the laboratory in order to eliminate possible measurement inaccuracies due to differences in techniques, environmental conditions, choice of higher level standards, etc. These calibration procedures may be prepared by the laboratory, or the laboratory may use published standard practices or written instructions provided by the manufacturer of the equipment. These procedures should include the following information.

1. The identification of the type of equipment for which the procedure is applicable, to include: Nomenclature, Model number or numbers, and type.
2. A brief abstract of the scope, principle, or theory of the calibration method.
3. A list of calibration standards and accessory equipment required to perform the calibration described. The manufacturer's name, model number, and all other pertinent information about accessory equipment should be provided.
4. The complete, detailed procedure for calibration, arranged in a step-by-step manner, clearly and concisely written.
5. Calibration procedures should provide specific instructions for obtaining and recording data, and should include a copy of any special forms necessary for recording data obtained during the calibration procedure.
6. Specification of requirements for statistical analysis of results if necessary.

GOVERNMENT-INDUSTRY DATA EXCHANGE PROGRAM (GIDEP)

Instrument and gage calibration procedures are often difficult to obtain, therefore, since the GIDEP Metrology Data Exchange function is a source of over 19,000 written calibration procedures, it will be brought to the attention of readers at this point.

The Government-Industry Data Exchange Program (GIDEP) is a government-sponsored program designed to facilitate the exchange of data among government activities and contractors. GIDEP, since its inception in 1960, has amply demonstrated a mutual benefit for participants with documented savings in excess of $160,000,000. The Metrology Data Exchange function became a part of the GIDEP Program in 1968. This data bank contains over 19,000 calibration procedures and metrology-related documents. Government facilities, prime contractors, subcontractors, manufacturers, and business firms involved in the use of, and calibration of, test instrumentation are currently participating in the Metrology Data Interchange. Information contained within the Metrology Data Bank includes calibration procedures, maintenance and repair manuals, specifications and standards, instrument rework procedures, measurement techniques, and other technical information related to the fabrication, application, and calibration of test and analytical instrumentation and gaging. The Metrology Data Interchange was established to reduce duplication of effort and costs expended by both the government and the private sector for the preparation of gage and instrument calibration procedures and related metrology data.

GIDEP operates under an agreement of the Joint Commanders of the Army Materiel Command (AMC), Naval Materiel Command (NMC), Air Force Logistics Command (AFLC), and Air Force Systems Command (AFSC). A Charter established the Program Manager Office within the MNC. A Government Advisory Group and an Industry Advisory Group act in an advisory capacity to the Program Manager. The GIDEP Administration office implements the functions of the GIDEP as directed by the Program Manager. The technical operation of the program is conducted through the Administration Office located at the Fleet Analysis Center, Corona, California.

The operation of the Metrology Data Bank is straightforward and simple. Gage and instrumentation equipment procedures and metrology-related documents prepared by participants in the program are submitted through an in-house Program Representative to the GIDEP Administration Office for processing. The Administration Office reviews, microfiches, reproduces, and distributes processed materials to all GIDEP participants. A microfiche data bank is furnished to each full-distribution status participant, thus providing immediate access to all information contained in the Metrology Data Bank. Descriptive indexes of all current calibration procedures are prepared and distributed by the GIDEP Administrtion Office. A remote terminal inquiry service which uses a time-share computer system is also available through GIDEP. In addition, an Urgent Data Request (UDR) system is available for participants to obtain metrology data which are unobtainable through other sources, but which possibly exist in other participating GIDEP organizations.

Participation in the GIDEP Metrology Data Interchange Program is voluntary. While GIDEP was originally established primarily for the use of government agencies and their contractors, others outside these bodies may participate, providing they meet the requirements outlined below. To apply for participation, a formal letter of request must be directed to the GIDEP Administration Office and must be signed by an official duly authorized to commit the organization to the obligation association with participant status. None of the direct costs of the program are assessed against participants.

The Metrology Data Bank is available on 24X microfiche. New acquisitions, revisions, and changes are issued whenever sufficient material accumulates. This data bank of over 19,000 calibration procedures and related information is furnished free of charge to qualified participants whose applications have been approved by the GIDEP Program management.

Applications for participation must be submitted on company stationery directed to the address given below. The basic admission requirements, discussed immediately following must be addressed point by point in the application letter.

The letter must request either full or partial distribution status. Full (*A*) distribution status is available to organizations whose level of metrology activity is great enough to justify custody of the microfiche data bank, including future releases. partial (*B*) distribution status is more suitable to those contractors and agencies whose need is less demanding, and who could obtain the desired metrology data by requesting the microfiche containing the needed data on a loan basis from the GIDEP Administration Office, rather than maintaining the entire microfiche data bank in-house. Copies of the desired data then can be made with the microfiche reader-printer, and the borrowed microfiche returned to the Administration Office.

The applicant, of course, must acquire or have access to a suitable microfiche reader-printer.

The applicant must agree to appoint a responsible person to act as GIDEP Metrology Data Interchange Representative. Suitable physical facilities and clerical assistance should be made available.

Initial submittal of at least one calibration procedure or metrology-related report reasonably representative of future submittals is required.

The letter should be directed to: GIDEP Operations Center, Corona, California, 91720; Phone: (714) 736-4677; Autovon: 933-4677.

All participants will be provided with a Policies and Procedures Manual and a complete set of Indexes to the Data Bank. In addition A-participants will receive the microfiche data bank.

CALIBRATION SOURCE

All calibrations performed by or for the laboratory should be traced through an unbroken chain, supported by reports or data sheets to some ultimate or national reference standards maintained by an organization such as the National Institute for Science and Technology (NIST). The ultimate reference standard can also be an independently reproducible standard, i.e., a standard that depends on accepted values of natural physical constants.

An up-to-date calibration report for each calibration standard used in the calibration system should be provided. If outside calibration services are performed on a contract basis, copies of reports issued should be maintained on file.

Copies of all calibration records (Figure 13-2) should be kept on file and should contain the following information:

- Description of the equipment
- Manufacturer of the equipment
- Model name, model number, and serial number
- Calibration frequency
- Calibration procedure number to be used
- Location of equipment
- Calibration date
- Calibration measurement obtained and corrected values
- By whom calibrated
- Assigned calibration interval

LABELING

All equipment in the calibration system should have, affixed to it in plain sight, a tag or label bearing the following information (see Figure 13-3):

- The date last Calibrated. _____
- Calibrated by what person. _____
- Next calibration due date _____

If the equipment size or its intended use limits or prohibits the use of a tag or label, an identifying code should be used.

Equipment past due for calibration should be removed from service either physically or, if this is impractical, should be impounded by tagging (Figure 13-4) or other means. Use of out-of-calibration equipment should be prohibited (see Figure 13-5).

References

10 June 1980. *MIL-STD-45662A, Calibration System Requirements.* Washington, D.C.: Department of Defense.

July, 1964. *Evaluation of a Contractor's Calibration System, MIL-HDBK 52.* Washington, D.C.: Department of Defense.

January 1975. *Collaborative Study of method 10—Reference Method for Determination of Carbon Monoxide from Stationary Sources.* Research Triangle Park, NC: EPA, National Environmental Research Center.

July, 1983. *Industrial Hygiene Laboratory Quality Control.* Cincinnati, OH: National Institute for Occupational Safety and Health.

SECTION 15 PREVENTIVE MAINTENANCE

As defined here, preventive maintenance is an orderly program of positive actions such as equipment cleaning, lubricating, reconditioning, adjustment, or testing to prevent instruments or equipment from failing during use. The most important effect a good preventive maintenance program has is to increase measurement system reliability, and thus increase data completeness. Conversely, a poor preventive maintenance program will result in increased measurement system downtime (i.e., a decrease in data completeness), increase maintenance costs, and may cause distrust in the validity of the data. Data completeness is one of the criteria used to validate data. See Section 17 for a discussion of data validation.

Laboratory managers should prepare and implement a preventive maintenance schedule for measurement systems. The planning required to prepare the preventive maintenance schedule will have the effect of: (1) Highlighting the equipment (or parts thereof) that is most likely to fail without proper preventive maintenance; and (2) Defining a spare parts inventory which should be maintained to replace worn-out parts with a minimum of downtime.

The laboratory preventive maintenance schedule should relate to the purpose of the analysis or test, environmental influences, the physical location of the equipment, and the level of operator skills. Checklists are commonly used to list required maintenance tasks and the frequency or time interval between scheduled maintenance operations.

When sampling includes several instruments, it becomes important to integrate checklists into the preventive maintenance schedule. Since instrument calibration is sometimes the responsibility of the operator, in addition to preventive maintenance, and since calibration tasks may be difficult to separate from preventive maintenance tasks, a combined preventive maintenance-calibration schedule may be appropriate.

A record of all preventive maintenance and daily service checks should be kept. Normally, it is convenient to file the daily service checklists with any measurement data. An acceptable practice to follow for recording task completion is to maintain a preventive maintenance-calibration multiple copy maintenance record log book. After tasks have been completed and entered in the instrument log book, a copy for each task is removed and sent to the supervisor for review and file. At the minimum, instrument logs will contain a record of the routine performance check results and maintenance done on the instrument, as well as a record of the day-to-day use of the instrument. The instrument log book should be clearly marked to show the instrument identification and should be kept near the instrument.

Reference

July, 1983. *Industrial Hygiene Laboratory Quality Control.* Cincinnati, OH: National Institute for Occupational Safety and Health.

SECTION 16 REFERENCE STANDARDS AND STANDARD REFERENCE MATERIALS

Since we have stated that all measurements should be based on calibration against reference standards or standard reference materials, it is incumbent upon laboratories to obtain reliable reference standards for calibration work. Such standards should be periodically checked against standards of higher accuracy, or against standard reference materials (see Section 14). This phase of internal quality control is critical for laboratories doing trace analytical work.

For each method, the analyst must estimate the approximate number and range of standards that will be necessary, using information gained from past experience, or that given by the method. The source of the standard should be determined. Standard Reference Materials (SRMs) from the National Institute for Science and Technology should be used whenever possible. All standard materials should be assayed to assure that they are of sufficient purity for the analysis being performed. The methods for performing this assay will vary, depending on the technique being used.

Over 1000 SRMs are available from the National Institute for Science and Technology. For price lists, ordering instructions, or for NIST Special Publication 260, NIST Standard Reference Material Catalog, contact: Office of Standard Reference Materials, Room B311, Chemistry Building, National Institute for Science and Technology, Washington, D.C. 20234; Telephone: 301-921-2045.

Also see: NIST Special Publication 250, Calibration and Related Measurement Services of the National Institute for Science and Technology, Available from Superintendent of Documents, U.S. Government Printing Office, Washington, D.C.

References

July, 1983. *Industrial Hygiene Laboratory Quality Control.* Cincinnati, OH: National Institute for Occupational Safety and Health.
January, 1987. *Quality Assurance and Laboratory Operations Manual.* Cincinnati, OH: National Institute for Occupational Safety and Health.

SECTION 17 DATA VALIDATION

Data validation is the process during which data are checked and accepted or rejected based on an established set of criteria. This requires the critical review of a body of data to locate and identify spurious results. It may involve only a cursory scan to detect extreme values, or spot outliers, or a detailed evaluation requiring the use of a computer. In either case, when a suspect value is located, it is not immediately rejected. Each questionable value must be checked for validity. Records of values that are judged to be invalid, or are otherwise suspicious, should be kept. These records are, among other things, useful sources of information for judging data quality. There are two methods of data validation, manual inspection and using computerized techniques.

When employing manual data validation, both the analyst or test technician and the laboratory supervisor should inspect integrated daily or weekly results for questionable values. This type of validation is most sensitive to extreme values, i.e., those which are higher or lower than expected values or which appear outside control limits. These latter are called "outlying observations" or, simply, "outliers."

The criteria for determining an extreme value are derived from prior data obtained from results of similar methods or, when necessary, by applying the appropriate statistical test to determine how to treat the outlying observation.

The time spent checking data that has been manually reduced by technicians depends on the time available and on the demonstrated abilities of the personnel involved.

The Environmental Protection Agency suggests an audit level of 7 percent, i.e., checking 7 out of every 100 values. This audit level is somewhat arbitrary, and it would be expected to be changed as more experience with the method is gained.

Computerized techniques can be used not only to retrieve, but also to validate data. The basic system for checking extreme values by manual techniques also applies here. However, the criteria for identifying extreme values may be refined in a number of ways to pinpoint suspect data as affected by various outside conditions. For instance, the program could be made to be specific for individual hours during a period of continuous monitoring. In this way, as an example, an hourly average concentration of carbon monoxide in an air sample taken at 8:00 AM, considered to be normal, would be flagged as abnormal if the same concentration appeared at 2:00 AM.

Another indication of spurious data that could be flagged for attention is a large difference in values reported for two successive time intervals. The difference in concentration values, which might be considered excessive, may vary from one contaminant to another and quite possibly may vary from one sampling location to another for the same contaminant. Ideally, this difference in concentration is determined through a statistical analysis of historical data. For example, it may be determined that a difference of 0.05 ppm in an SO_2 concentration for successive hourly averages occurs rarely (less than 5 percent of the time). But at the same location, the hourly average CO concentration may change by as much as 10 ppm. The criteria for what constitutes an excessive change may also be linked to the time of day and contaminant relationships, e.g., high concentrations of SO_2 and O_3 cannot coexist, and data in which this occurs should be considered suspect.

While the examples above deal with industrial hygiene data, the principles illustrated are valid for application in many laboratory situations outside the field of industrial hygiene.

The validation criteria for any data set should ultimately be determined by the objectives for collecting the data. The extent of the decision elements to be used in data validation cannot be defined for the general case. Rather, the validation criteria should be tailored along the lines suggested earlier for varying types of contaminant determinations.

There are several statistical tools that can be used in the validation of data generated by continuous monitoring strip charts displaying an analog trace. Usually strip charts are cut at weekly intervals and are turned over to data-handling staff for interpretation. The technician may estimate by inspection the hourly average contaminant concentrations and convert the analog percent of scale to other units, for example, ppm.

Reading strip charts is a tedious job subject to varying degrees of error. A procedure for maintaining a desirable quality for data manually reduced from strip charts is important. One procedure for checking the validity of the data reduced by one technician is to have another technician or supervisor check the data. Because the values have been taken from the chart by visual inspection, some difference in the values derived by two different individuals can be expected. When the difference exceeds a specified amount and the initial reading has been determined to be incorrect, an error should be noted. If the number of errors exceeds a predetermined number, all data for that strip chart are rejected and the chart is read again by an individual other than the one who originally read the chart. The question of how many values to check can be answered by applying acceptance sampling techniques, such as the use of ANSI/ASQC Standard Z-1.4, "Sampling Procedures and Tables for Inspection by Attributes."

Acceptance sampling can be applied to data validation to determine the number of data items (individual values on a strip chart) that need to be checked to determine, with a given confidence level, that all data items are acceptable. Management wants to know, without the necessity of checking every data point, if a defined error level has been exceeded. From each strip chart with N data values, the supervising checker can randomly inspect n data values. If the number of erroneous values in less than or equal to c, the rejection criterion, the values for the strip chart are accepted. If the number of errors is greater than c, the values for the strip chart are rejected, and another individual is asked to read the chart. An explanation of how to determine

sample, acceptance, and rejection values appears in ANSI/ASQC Standard Z-1.4.

Another useful technique in determining the validity of data is the use of statistical tests for the significance of difference in data. In this approach, the collection of data from a sample of fixed size is required. A statistic is then computed from the sample and compared with critical values given in the appropriate tables for the test selected. Examples are the t-test, X^2- (chi-square) test, F-test, and so forth. Discussions of the proper applications of these and similar tests for the significance of differences in data can be found in standard statistics texts. When using such a procedure, it is necessary to collect the specified sample observation regardless of the results that may be obtained from the first few observations.

A procedure called sequential analysis requires that a decision be made after each observation or group of observations. This procedure has the advantage that, on the average, a decision as to the acceptability of the data can be reached with fewer observations than a fixed sample size requires. For a discussion of the use of sequential sampling plans, see Duncan's *Quality Control and Industrial Statistics*.

References

1967. *MIL-STD-781B. Reliability Tests, Exponential Distribution.* Washington, D.C.: Department of Defense.

Burr, Irving W. 1953. *Engineering Statistics and Quality Control.* New York; McGraw-Hill Book Co.

July 1983. *Industrial Hygiene Laboratory Quality Control.* Cincinnati, OH: National Institute for Occupational Safety and Health.

1981. *ANSI/ASQC Z1.4—1981 American National Standard. Sampling Procedures and Tables for Inspection by Attributes.* Milwaukee, WI: American Society for Quality Control.

SECTION 18 ENVIRONMENTAL CONTROLS

Aside from the required control of environmental conditions for calibration activity discussed in Section 14 (see page 12), it may often be necessary for the laboratory to control the atmospheric and other working conditions for all laboratory operations to the extent necessary to ensure the precision and accuracy of laboratory results. Where the test or method spells out special ambient conditions for the conduct of the test or analysis, then the measures taken to control environmental conditions should be described and the resulting data defining working conditions recorded. Such activities include not only elements such as those discussed in

Section 14, but also such things as restricted access provisions, clean room operations (including Standing Operating Procedures), and special housekeeping and safety practices.

Reference

1987. *ANSI/ASQC Standard M1-1987. American National Standard for Calibration Systems.* Milwaukee, WI: American Society for Quality Control.

SECTION 19 CUSTOMER COMPLAINTS

The laboratory should have a formal, established procedure for handling technical questions and complaints, whether they originate with customers or regulatory or accrediting bodies. One individual should be assigned the responsibility of handling such inquiries or complaints. The duties of this individual include, in addition to answering the inquiry or complaint:

• Circulating information as to the nature of the complaint to all interested personnel within the laboratory.
• Conducting a preliminary investigation to determine the nature and validity of the complaint.
• Initiating a request for corrective action if necessary.
• Advising management if the nature of the complaint is serious or might lead to legal action.
• Preparing periodic reports to management with regard to the frequency and status of inquiries and complaints.

Reference

Juran, J. M., 1974. *Quality Control Handbook*, 3rd Ed. New York: McGraw-Hill Book Company.

SECTION 20 SUBCONTRACTING

It sometimes occurs that a laboratory may not have the expertise or equipment necessary to conduct a certain test or analysis when it is one of a battery of tests or analyses that the company is asked to perform. In such cases, the laboratory may elect to have the work performed at an outside laboratory which is competent to perform the task required. In spite of the fact that the work will be done by others, the laboratory purchasing or contracting for the work bears the ultimate responsibility for the quality of the work produced, and is responsible for assuring that all services procured from an outside source conform to requirements of the original customer. The selection of a subcontracting laboratory and the nature and extent of surveillance or control should be dependent on the nature of the test or analy-

sis to be performed, that is, how difficult or sensitive it is; the subcontractor's demonstrated ability to perform, as witnessed by being accredited, or by records of past performance; and the quality evidence made available. To assure an adequate and economical control of the quality of the results, the contracting laboratory should use, to the fullest extent possible, objective evidence of quality furnished by the subcontractor, such as data validation results, calibration curves, and results of routine quality checks such as blank determinations and results of duplicate or replicate tests or analyses.

It may be necessary to determine the effectiveness and integrity of the control of quality by an outside laboratory by assessment and review at intervals consistent with the complexity and degree of usage of the service required. When conducting such assessments or audits, by test or analytical means, all available objective evidence relating to the source's control of quality should be used in conducting the audit of the subcontracting laboratory. (Section 26).

The laboratory's responsibility for the control of purchased services includes the establishment of procedures for:

1. The selection of qualified outside laboratories.
2. The transmission of applicable technical and/or method requirements.
3. The evaluation of the test and analytical reports received, before providing them to the customer.
4. Providing effective provisions for early information feedback and correction of nonconformances.

In summary, when a laboratory contracts work to other, outside laboratories, it must endeavor to ensure that the work contracted for meets the quality standards required by its own customer and must take steps accordingly.

Reference

31 October 1960. *Handbook H50 Evaluation of a Contractor's Quality Program.* Washington, D.C.: Department of Defense.

SECTION 21 PERSONNEL

QUALITY TRAINING

All personnel involved in any function affecting data quality (sample collection, analysis, data reduction, calibration of instruments, and other quality assurance activities) should have sufficient training in their appointed job to enable them to generate and report accurate, precise, and complete data. The Quality Control Coordinator should bear the responsibility for seeing that the required training is available for these personnel, and for taking action when it is not.

Quality Control training programs should have the objective of seeking solutions to laboratory quality problems. This training objective should be concerned with the development, for all laboratory personnel involved in any aspect or function affecting quality, of those attitudes, that knowledge, and those skills which will enable each one to contribute to the production of high quality data continuously and effectively.

A number of training methods for laboratory personnel dealing with quality control are available.

(1). Experience training or "On-the-Job" (OJT) training is the process of learning to cope with problems using prior experience or knowledge as the basis for action.

(2). Guidance training is OJT with outside help from supervisors or co-workers. The advice may be solicited or provided on an informal basis, or on a planned, structured basis.

Employees involved in an effective program employing OJT techniques will:

(a) Observe experienced technicians or operators perform the necessary steps in a test or analytical method.

(b) Perform the various operations in the method under the supervision of an experienced technician or operator.

(c) Perform operations independently but with a high level of quality control checks utilizing the proficiency evaluation procedures discussed in Section 12.

3. Be engaged in independent study involving attendance at night school classes, outside reading, attendance at seminars, or taking correspondence courses on a voluntary basis.

4. Attend in-house training, classroom study taken during working hours, presented on a formal basis. Such classes may be presented as short courses lasting from two or three days to two weeks, on general or specialized subjects. Numerous universities and technical schools offer long-term, quarter, and semester-length academic courses in statistics, computer technology, and other courses of interest to the quality technologist.

QUALIFICATION RECORDS

Certain complex testing or analytical techniques or instruments may require specialized training and the formal qualification of technicians or operators in the performance of such specialties. Once individuals are qualified, records must be kept and requalification undertaken periodically, as necessary. The Quality Control Coordinator should be made responsible for the maintenance of such records and for seeing that requalification is accomplished in a timely manner.

TRAINING EVALUATION

Evaluation of the effectiveness of training is accomplished by the conduct of periodic intralaboratory proficiency testing of laboratory personnel involved in the conduct of testing or analytical activity. This evaluation should lead to the determination of the level of knowledge and skill achieved by the technician from the training and appraisal of the overall effectiveness of the training effort, including a finding of any training areas that need improvement.

MOTIVATION

The incentive to produce results with consistently high quality must be provided from the top of the laboratory organization. As a management policy, the concept of "Total Quality Control" may be embraced by management. Feigenbaum describes "Total Quality Control" as follows:

Total quality control is an effective system for integrating the quality-development, quality-maintenance, and quality improvement effort of the various groups in an organization so as to enable production and service at the most economical levels which allow for full customer satisfaction.

Even though many individuals and smaller elements of a laboratory organization may be involved in quality control activities, the ultimate responsibility for driving the quality effort rests on the shoulders of top management. The best way for a laboratory director or supervisor to demonstrate his desire to achieve high quality results is to show continuous, conscientious, participatory interest in quality activities. Some of the methods of initiating such things are listed below.

Motivation by Communication

(a) Emphasizing quality-related articles in company newsletters.

(b) Quality bulletins.

(c) Quality Propaganda Posters. These are available from a number of commercial sources.

(d) "Horror story" displays of major quality accidents, taking care to keep the participants anonymous and unidentifiable.

(e) Public award ceremonies for good work.

(f) Open house programs to demonstrate to employees, customers, employees' families, and the public how the laboratory operates, stressing quality efforts.

Motivation Campaigns

Formal motivational campaigns are effective only if they are:

(a) Well planned and organized.
(b) Have a specific goal or aim.
(c) Are finite, that is, they must have a well defined starting point, a planned, well thought-out path of activity, and an identifiable termination point.

Examples of such motivational campaigns which have had varying degrees of success are: the Zero Defects programs, the Polish DO:RO System, the Soviet Saratov System, and the Quality Circle Movement which was introduced in Japan by Deming and Juran in 1962 and has been used extensively and successfully in that country. For more information on these programs and on reference material regarding them, see Juran's *Quality Assurance Handbook*, 3rd Edition, Chapter 21.

References

Feigenbaum, A. V. 1961. *Total Quality Control.* New York: McGraw-Hill Book Company.

Feigenbaum, A. V. 1954. "Company Education in the Quality Problem." *Industrial Quality Control.* **X** (6):24–29.

Juran, J. M. 1962. *Quality Control Handbook.* 2nd Ed. New York: McGraw-Hill Book Company.

Juran, J. M. 1974. *Quality Control Handbook,* 3rd Edition. New York: McGraw-Hill Book Company.

Reynolds, E. A. 1954. "Industrial Training of Quality Engineers and Supervisors." *Industrial Quality Control.* **X** (6):13–20.

Reynolds, E. A. 1970. "Training QC Engineers and Managers." *Quality Progress.* **III** (4):20–21.

1967. *Industrial Quality Control.* **XXIII** (12). All articles deal with Quality Education and Training.

1983. *Industrial Hygiene Laboratory Quality Control.* Cincinnati, OH: National Institute for Occupational Safety and Health.

SECTION 22   STATISTICAL METHODS

There are a number of statistical tools available to the laboratory practitioner which can be used to obtain more information about the data produced from analytical or test results. The purpose of this discussion is to provide a brief description of statistical techniques that are most used and are most useful in the laboratory, without presenting mathematical details. For further study, the reader is invited to the reference material given at the end of this section, or to any of the standard text books on statistics.

The control chart is perhaps the most useful and most commonly used statistical tool available to the laboratory. The use of control charts on a routine basis is also a requirement of many accreditation programs.

The control chart provides a method for distinguishing the pattern of intermediate (random) error or variation from the determinate (assignable cause) error or variation. This technique displays the test or analytical data from the test or analysis in a form which graphically compares the variability of all test results with the average or expected variability of small groups of data —in effect, a graphic analysis of variance, which is a comparison of the "within groups" variability versus the "between groups" variability.

The test or analytical data are plotted on a chart in units of the test result on the vertical scale, against time or the sequence of tests on the horizontal scale. After a number of data points have been plotted, preferably not less than twenty, the average or mean value is calculated and plotted on the chart. The data performance can then be compared to the mean of the data group. There are many forms of control charts, but the type most commonly used in laboratories is the $\overline{X}$-R chart. In this type of chart, the average of a group or sample of tests is plotted on an $\overline{X}$ Chart and compared with the grand mean or $\overline{\overline{X}}$ "average of averages" of the data. Additionally, the upper and lower control limits are calculated and plotted at three standard deviations above and below the grand mean. At the same time, the sample ranges are plotted on a companion $\overline{R}$ chart, and the average range $\overline{R}$ and the upper and lower control limits on the range and calculated and plotted in a similar manner. (Figure 22-1). Data points which fall outside the upper and lower control limits indicate a possible out-of-control condition.

The laboratory should decide which control charts it intends to use, where they are to be used, who is responsible for establishing and maintaining the charts. Once these decisions are made, procedures should be put in place to formalize the use of control charts as a part of the laboratory's routine work.

An unusually large or small value or measurement in a set of observations is referred to as an "outlier" in the statistical literature. Outliers are particularly interesting to the laboratory technician because they may appear as data points lying outside the upper or lower control limit in a control chart. The question of testing whether or not the appearance of the anomaly is of any significance then becomes important. The purpose of such tests would be to:

1. Screen data for outliers, and identify the need for closer control of the data-generating process.
2. Eliminate outliers prior to analysis of the data. For example, in the development of the control chart, the presence of outliers would lead to limits which are too wide and would make the interpretation and use of the control chart invalid. Incorrect conclusions are likely to result if the outliers are not eliminated prior to analysis of the data.

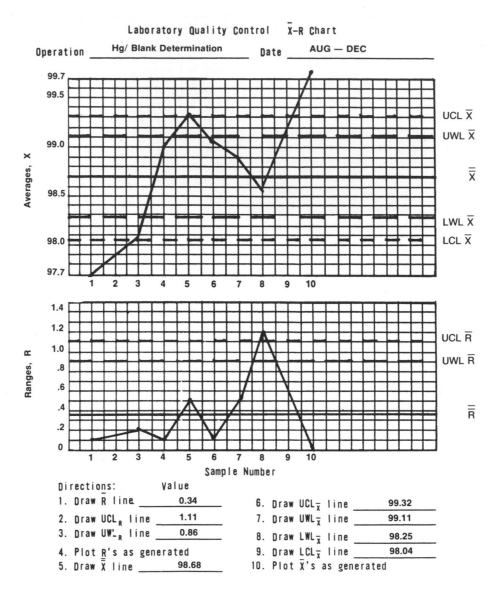

Figure 22-1

3. Identify the outliers which occur due to unusual or changing conditions of measurement, for example, a carbon monoxide (CO) concentration which is abnormally large due to a local change in environmental conditions during the time of sample collection. Such observations would not be indicative of the average concentration of CO and may be eliminated depending on the use of the data. Ideally, these unusual conditions should be recorded on the field data report.

In the laboratory, the need often arises to determine whether or not the differences in sets of data are statistically significant. There are a number of tests available to enable the user to decide upon the significance of such differences when they occur. The selection of the proper statistical test depends upon the amount and

kind of information that is available concerning the results in question. Three tests in common use for determining the significance of differences are: the t-test, the F-test, and the chi-square (X^2) test. All of these are based on different patterns of probabilities of distribution. The tables and equations related to these distributions may be found in available text books.

The t-test may be used for the determination of the significance of a difference between the mean or average value of a sample compared with the mean of the population when the population standard deviation is known, and, when the population standard deviation is unknown, but can be estimated from the sample standard deviation. The t-test is also used for the determination of the significance of differences between the average or mean values of two different samples drawn from the same population, when the population standard deviation is known, or when the population standard deviation is unknown and is believed to be the same for both samples.

The chi-square test is used to determine the significance of differences between the sample standard deviation and the population standard deviation, when the population standard deviation is known.

The F-test is used to determine the significance of a difference between two different sample standard deviations when the population standard deviation is unknown.

The laboratory may need to use sampling techniques in the conduct of data validation, as we have seen in Section 17. In addition, certain larger organizations, which buy laboratory supplies in large quantities, may find it necessary to verify the quality of incoming lots of such material in order to ensure that it meets specified standards. In such cases, the use of sampling plans during incoming inspection is advisable as an economic alternative to 100 percent inspection of the materials.

For simple, nondestructive inspection procedures, the use of acceptance sampling by attributes is customary. This involves determining whether a lot of items should be accepted on the basis of given specifications. A defective or nonconforming item which has a physical attribute which falls outside specified limits is considered to be a "reject." Thus, the sampling is by attributes, for instance, an item is identified as either a defect or a good item. This is often referred to as go-no-go inspection, where this refers to whether the item meets the specification when checked by a gage, or other measuring device, or is inspected visually. A tabulation of sampling plans, with a complete discussion of the employment of such plans, appears in ANSI/ASQC STD Z1.4-1981, listed in the references below.

Where inspection or testing is destructive, time-consuming, or expensive, a considerable savings in sampling may be achieved if the decisions concerning the acceptance of a lot or data set can be made on the basis of the actual measurements (a continuous value) rather than whether the measurements are outside specified limits. The decision on acceptance of a lot is made by variables, that is, on the basis of the comparison of the mean and standard deviation of the sample of n measurements and given constants taken from tables provided. Variable sampling plans are described and the related tables are furnished in ANSI/ASQC STD Z1.9-1981 referenced below.

References

1983. *Industrial Hygiene Laboratory Quality Control.* Cincinnati, OH: National Institute for Occupational Safety and Health.

1975. *Quality Assurance Handbook for Air Pollution Measurement Systems, Vol. I Principles.* Research Triangle Park, NC: U.S. Environmental Protection Agency.

1980. *American National Standard—Sampling Procedures and Tables for Inspection by Variables for Percent Nonconforming ANSI/ASQC Z1.9-1980.* Milwaukee, WI: American Society for Quality Control.

1981. *American National Standard—Sampling Procedures and Tables for Inspection by Attributes-ANSI/ASQC Z1.4-1981.* Milwaukee, WI: American Society for Quality Control.

SECTION 23 NONCONFORMITY

American National Standard ANSI/ASQC Z1.4-1981 defines "nonconformity" as:

> A departure of a quality characteristic from its intended level or state that occurs with a severity sufficient to cause an associated product or service not to meet a specification requirement. (Underline by the author.)

The question to be resolved is, what action should the laboratory take when, as a result of test or analytical procedures, customer feed-back, management audits, or data validation, nonconforming results are encountered?

The laboratory should establish a set of procedures to be followed when nonconforming results appear, whether obvious or suspected. Such procedures should incorporate, but are not limited to:

1. Reporting and recording the occurrence of a nonconforming event.
2. Suspension of work, and commencement of an investigation.
3. Report to the customer, if necessary.
4. Repetition of the test or analysis.
5. Initiate corrective action to prevent reoccurrence of the nonconformance.

SECTION 24 CORRECTIVE ACTION

In a quality assurance program, one of the most effective means of preventing trouble is to respond immediately to reports of suspicious data or equipment malfunctions from the test or analytical operator. The application of proper corrective action at this point can reduce or prevent the production of poor quality data. Established procedures for corrective action are often included in the method for the operator's use when performance limits are found to be exceeded, either through direct observation of the parameter in question, or through review of control charts. Specific control procedures, calibration, presampling, or preanalysis operational checks, and so forth, are designed to detect instances in which corrective action is necessary. A checklist for logical alternatives for tracing the source of a sampling or analytical error is provided to the operator.

Troubleshooting guides for operators, field technicians, or laboratory test or analytical technicians are generally found in instrument manufacturer's manuals. On-the-spot corrective actions routinely made by technicians should be documented as normal operating procedures, and no specific documentation other than notations in the laboratory workbook need be made.

Long-term corrective action is taken to identify and permanently eliminate causes of repetitive nonconformance. To improve the quality of test results to an acceptable level and to maintain that quality at that level, it is necessary that the quality assurance system be sensitive and timely in detecting out-of-control or unsatisfactory conditions. It is equally important that, once the conditions of unacceptable results are indicated, a systematic and timely mechanism is established to assure that the condition is reported to those who can correct it and that a positive loop mechanism is established to assure that appropriate corrective action has been taken in a timely manner. For major problems, it is desirable that a formal system of reporting and recording of corrective actions be established.

Experience has shown that most problems will not disappear until positive action has been taken by management. The significant characteristic of any good management system is the step that closes the loop—the initiation of a change if the system demands it.

Effective corrective action occurs when several individuals and departments cooperate in a planned program. There are several essential steps that must be taken to implement a corrective action program that achieves desired results. These steps form the closed-loop system that is necessary for corrective action success, to wit:

1. Method specifies the required quality.
2. The Quality Report compares actual results with expected specified results and reports a nonconformity.
   (Note: This is a method, test, or analytical nonconformity—not a nonconformance to specification)
3. The Corrective Action Analyst, who has been given responsibility for the task, initiates an investigation, research, engineering, testing activities which result in necessary changes, completing this work in a timely manner, meeting specified deadlines.
4. The Corrective Action Analyst reports on corrective action measures taken.
5. The Quality Control Coordinator follows the method, checks and analyzes quality to ensure that the corrective action "fix" is appropriate and has succeeded in achieving the results desired.

Corrective actions should be a continual part of the laboratory system for quality, and they should be formally documented. Corrective action is not complete until it is demonstrated that the action has effectively and permanently corrected the problem. Diligent follow-up is probably the most important element of a successful corrective action system.

The document usually employed in a Corrective Action System is the Corrective Action Request (Figure 23-1, Part 4) or CAR. The CAR may be initiated by any individual in the laboratory who observes a major problem. The CAR should be limited to a single problem. If more than one problem is involved, each should be documented on a separate corrective action request form.

Corrective action can be informal if the organization is small or the problems few. When this is not the case, and the problems are severe or numerous, corrective action status records may be needed. In addition to the CAR, the system is supplemented by the use of a Corrective Action Master Log (Figure 23-2, Part 4). Each CAR is assigned a sequential number and logged in with the appropriate information. Even if a problem is reported by letter, memo, customer complaint, or by other means, a CAR should be completed and logged in for action and follow-up.

Reference

1983. *Industrial Hygiene Laboratory Quality Control.* Cincinnati, OH: National Institute for Occupational Safety and Health.

SECTION 25 QUALITY COST REPORTING

As a management tool, quality assurance costs should be identified and recorded primarily to identify elements of quality assurance programs whose costs may be disproportionate to the benefits derived. An additional purpose is to detect cost trends for budget forecasting.

Prior to identifying and setting up a quality cost system in the laboratory, the costs of the quality function are probably widely scattered within the cost accounting system of the organization. It is important then, to define those elements and subelements which go to make up the total quality cost package and adjust the cost accounting system so as to be able to accumulate and present an accurate quality cost picture to management.

The American Society for Quality Control lists four principal categories for quality costs, as follows:

1. *Prevention costs* associated with personnel engaged in designing, implementing, and maintaining the quality system. (Maintaining the quality system includes auditing the system.)
2. *Appraisal costs* associated with the measuring, evaluating, or auditing of products, components, and

purchased materials to assure conformance with quality standards and performance requirements.

3. *Internal-failure costs* associated with the manufacture of defective products, components, and materials that fail to meet quality requirements and thus result in manufacturing losses.

4. *External-failure costs* generated as a result of defective products having been shipped to customers.

An examination of these categories, which do not fit very well into the laboratory environment, will, however, give a frame of reference around which a new set of categories, attuned to laboratory operations, may be developed. Since, obviously, the quality assurance activities in laboratories will be different from those in manufacturing, it is more practical to categorize cost areas as follows:

1. *Prevention costs* associated with keeping unacceptable data from being generated in the first place. Included here are such things as quality, planning, quality control training, etc.

2. *Appraisal costs* associated with efforts to maintain measurement system performance quality levels by formal audits of performance quality levels and interlaboratory and intralaboratory testing programs, etc.

3. *Internal-failure costs* caused by the occurrence of determinations or test results that do not meet acceptance standards. These include voided data, spoiled test or analytical samples, and repeated or duplicated tests.

4. *External-failure costs* caused by unacceptable test or analytical results that have already left the laboratory. This involves effort spent in corrective action, investigations, and repeated tests or analyses required in order to gain customer satisfaction.

Within each major cost category appear a number of subelements, as follows:

1. *Prevention costs*
   a. Quality planning.
   b. Document control and revision, including measurement method write ups.
   c. Quality training.
   d. Quality assurance plans for projects and programs.
   e. The Quality Assurance Manual.
   f. Preventive maintenance.
2. *Appraisal costs*
   a. Quality assurance activities associated with pretest preparation, sample collection, sample analysis, and data reporting.
   b. Data validation.
   c. Procurement quality control.
   d. Statistical analysis of data.
   e. Calibration.

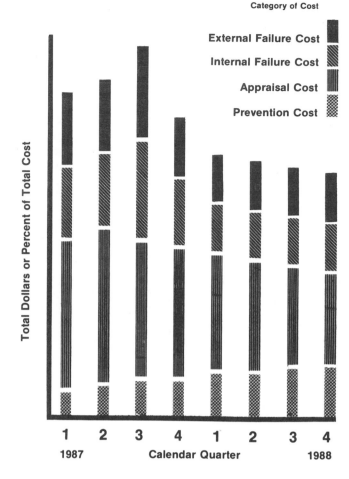

Figure 24-1

f. Interlaboratory and intralaboratory testing.
g. System audits.
h. Quality reports to management, including quality cost reports.
3. *Internal-failure costs*
   a. Scrapping of defective materials.
   b. Cost of rerunning tests or repeating analyses.
   c. Costs of corrective action efforts.
   d. Investigation or research efforts.
4. *External-failure costs*
   a. Investigation of complaints from outside sources.
   b. Cost of corrective action efforts.
   c. Cost of replacing samples and rerunning tests or analyses.

After quality cost elements have been listed, it is necessary to allocate to them cost figures available from the accounting system. If actual cost figures are not available, then the accounting component should provide estimates that are as nearly exact as possible.

It will usually be found that the costs are not uniformly distributed over the range of elements, a disproportionate percentage generally appearing in Appraisal and Internal- and External-failure costs. It has been found that a relatively small increase in Prevention expenditures will yield large reductions in Appraisal and Failure costs. Therefore, the laboratory management should consider adoption of preventive measures in order to reduce total quality control costs.

The quality control cost report should be presented periodically to management, showing the costs allocated to each of the four major quality cost categories and the relationships of the individual category costs to total quality costs. It may also be useful to provide further detail showing quality costs for each sub-element within the major cost categories. Graphic presentations should be made where appropriate (Figure 24-1).

References

Feigenbaum, A. V. 1961. *Total Quality Control. New York:* McGraw-Hill Book Company.

Juran, J. M. 1974. *Quality Control Handbook.* New York: McGraw-Hill Book Company.

Juran, J. M. and Gryna, F. M. 1970. *Quality Planning and Analysis.* New York: McGraw-Hill Book Company.

Masser, W. J. 1957. "The Quality Manager and Quality Costs." *Industrial Quality Control* **XIV** (4):5–8.

Rhodes, R. C. 1972. "Implementing a Quality Cost System." *Quality Progress* **5** (2):16–19.

1986. *Principles of Quality Costs.* Milwaukee, WI: American Society for Quality Control.

1983. *Industrial Hygiene Laboratory Quality Control.* Cincinnati, OH: National Institute for Occupational Safety and Health.

SECTION 26 QUALITY AUDITS

Quality audits* conducted within the laboratory fall into two general categories: performance audits and quality system audits.

Performance audits refer to independent checks made by a supervisor or auditor* to evaluate the quality of data produced by the sampling and testing or analytical system. Performance audits generally may be categorized as follows:

1. Sampling audits.
2. Analysis or test audits.
3. Data processing audits.

These audits are independent of, and in addition to, normal quality control checks by the operator, test technician, or analyst. Independence can be achieved by having the audit made by a different operator/technician/analyst than the one conducting the routine

---

*Some accrediting organizations may use the terms: *survey, assessment,* or *site visit* and *surveyor, assessor,* or *site visitor* instead of Audit and Auditor. For the sake of clarity we will use *audit* and *auditor* in this work.

measurements or, in the case of sampling or analysis, by the introduction of audit control standards into the sampling, testing, or analytical system, and the subsequent plotting of results on control charts by the supervisor. The use of audit control standards should be applied without the knowledge of the operator/technician/analyst, if possible, to insure that recorded results reflect normal operating conditions. Examples of performance auditing procedures are listed below.

1. Sampling audit. As an example, the auditor uses a separate set of calibrated flowmeters and reference standards to check the sample collection system using:
   a. Flow rate devices
   b. Instrument calibration
   c. Instrument calibration gases, when applicable.
2. Analysis or test audits. The auditor is commonly provided with set of a duplicate samples, or a split portion or aliquot of several routine samples for check analysis or test.
3. Data processing audits. Data reporting commonly involves a spot check on calculations, and data may also be checked by inserting in the data processing system a dummy set of raw data followed by review of the validated data.

A major challenge in audit planning is determining the audit frequency and, when dealing with data packages and reports in large numbers, the lot size, in order to determine the number of samples required to estimate population quality with a specified confidence level.

A system audit is an on-site inspection and review of the quality assurance system. Since most quality systems are described by the organization's quality manual, the audit becomes a check to see whether the laboratory is following all of the policies and procedures prescribed by its own manual. This not only involves a review of operating procedures, but a physical inspection of records such as Document Change Notices, Training Records, Calibration Records, etc.

System audits should be conducted by someone outside of the quality assurance activity in the laboratory, such as a supervisor from another laboratory within the company, the laboratory manager, an outside consultant, an auditor from an accrediting body, or, in some cases, an individual from the accounting component of the organization who is familiar with the overall operation of the laboratory.

Internal quality system audits should be conducted at least annually, and more often if conditions warrant. Provision should be made for follow-up audits to ensure that action has been taken to correct any deficiencies found.

Audit check lists should be prepared and used to minimize the possibility of overlooking any detail during the course of the audit. These check lists will provide a means of grading each of the various areas or quality elements checked during the audit, in order to pinpoint areas where need for improvement is found (Figure 25-1, Part 3).

For detailed information with regard to the conduct of an audit see ANSI/ASQC Q1-1986.

References

1980. *Nuclear Quality Systems Auditor Training Handbook.* Milwaukee, WI: American Society for Quality Control.

1986. ANSI/ASQC Q1-1986-American National Standard—Generic Guidelines for Auditing of Quality Systems. Milwaukee, WI: American Society for Quality Control.

1983. *Industrial Hygiene Laboratory Quality Control.* Cincinnati, OH: National Institute for Occupational Safety and Health.

SECTION 27 RELIABILITY

Some laboratories engaged in work where readings generated by remote recording instruments provide the data with which the laboratory must work are vitally interested in the reliability of such instrumentation. This section, then, is provided to offer the Quality Control Coordinator basic information about how to deal with concerns about instrument reliability.

The reliability of a measurement system is defined as the probability of the system performing its intended function for a prescribed period of time under the operating conditions specified. Conversely, unreliability is the probability of a device failing to perform as specified. The consideration of reliability is becoming increasingly important because of the increase in complexity and sophistication of sampling, analysis, automatic recording, and telemetering systems. Furthermore, data interpretation depends on data completeness (no breaks in the data set) for trend analysis. Generally, as equipment becomes more complicated, its probability of failure increases. The subject of reliability is complex. It is a discipline related to, but separate from, quality control. Therefore, reliability really deserves a separate, more detailed treatment. However, because of the relationship between reliability and quality control, a brief treatment of reliability elements is given here for the benefit of those laboratories for which unattended instrument performance is a matter of concern.

Quality assurance may be thought of as the activity monitoring product or service quality up to or at a given point in time, while reliability assurance is that activity concerned with satisfactory product or service performance over a specified or expected period of time.

In order to assure high reliability (i.e., completeness of data), several factors will need to be considered.

EXTERNAL FACTORS INFLUENCE RELIABILITY ASSURANCE

Reliability assurance systems are applied to a wide range of instruments, gages, and measuring systems. Differences are in complexity, kind of use, and the number of equipments involved. Factors that affect reliability or effectiveness, and which should be taken into consideration in planning programs are:

1. The severity of the reliability requirement.
2. Criticality of the device to the achievement of the desired results.
3. Type of equipment (continuous duty, intermittent duty, one-shot, etc.)
4. Ambient working conditions.
5. Established design or state-of-the-art equipment to be used.
6. Training, experience, and capability of the operating personnel.

RELIABILITY PROGRAMS MUST BE PLANNED IN ADVANCE

Reliability assurance programs, having been designed in consideration of the above factors, should be planned prior to implementation, to include the following tasks:

1. Specifications development.
2. Environmental requirements review.
3. Statistical test planning and test data reduction.
4. Failure data system planning.
5. Data analysis planning and programming.
6. Failed parts analysis.
7. Life tests to failure.
8. Quality control coordination.
9. Field tests.
10. Reliability prediction data collection.
11. Development of an organization responsible for the reliability function.

EQUIPMENT RELIABILITY REQUIREMENTS MUST BE SPECIFIED IN CONTRACTS

Equipment reliability requirements must be included as a specification in procurement contracts and must be met by the manufacturer. This specification should consist of:

1. The product reliability definition, which includes: (a) All functional requirements of the device; (b) Safety requirements; and (3) Resistance to environmental effects.
2. Allowable failure probabilities expressed as minimum mean time between failures (MTBF) or mean time to failure (MTTF).

Equipment Failure Rates MTBF or MTTF Can Be Predicted on Performance History

1. Reliability predictions are made through the summation of known or estimated failure rates of individual components which make up the system.
2. Estimates of component failure rates are obtainable in Military Standards and from the GIDEP Data Bank (see Section 14).
3. Estimates of the chance of failure-free operation for a specified time period are calculated, assuming the failure rate is constant, by:

$$P_s = R = e^{-t/m} = e^{-t\lambda}$$

   where:

   P_s = R = Probability of failure-free operation for a time period $\leq$ t.
   e = 2.718 (the natural log)
   t = A specified time of failure-free operation
   m = MTBF
   λ = Failure rate (reciprocal of m)
4. Mean time to failure rates will be estimated by the same formula, except that here, m now refers to MTTF.
5. Uptime will be the running-time data upon which the MTBF and MTTF records are based. Downtime will be the time span caused by failure between stop and start records or uptime subtracted from total elapsed running time.
6. Excessive downtime should be analyzed and corrective action taken, changing availability factors as necessary to improve operations. Such factors are:
   a. Redundancy
   b. Factors associated with spares
   c. Operational and maintenance factors
   d. Logistics support factors
   e. Detection capabilities.
7. Failure rate estimates should be made as data become available, and should be maintained in a current condition for each item and/or class of equipment. These estimates are made in order to:
   a. Estimate spares requirements.
   b. Estimate end item replacement requirements.
   c. Estimate end item backup requirements.
   d. Estimate maintenance requirements.
   e. Establish reliability goals.
   f. Compare realized MTBF values with estimates.
   g. Establish focal points for corrective action.

Incoming Equipment Should Be Tested and Inspected for Qualification and Adherence to Contract Specification for Reliability

1. Quality control acceptance tests should be conducted to determine whether the product in question meets performance and design specifications at the time of testing.
2. Reliability tests should be conducted to determine whether there is a high probability that the product will continue to meet the specified performance requirements, with the specified reliability, for its specified service life.
3. Qualification tests should be conducted on a sample or samples if feasible, testing to failure, to:
   a. Verify adherence to specified reliability standards.
   b. Generate data for product improvement.
   c. Provide an estimate of product service life and reliability.

Burn-in tests should be conducted for specified times where there is an indication of early failures.

Control the Operating Conditions

Environmental factors affecting performance or reliability may be natural, induced, or a combination of both.

1. Natural environmental factors are:
   a. Barometric pressure changes.
   b. Temperature.
   c. Particulate matter, such as sand, dust, insects, fungus, etc.
   d. Moisture, such as icing, salt spray, high humidity, etc.
2. Induced factors are:
   a. Temperature, self-generated or generated by adjacent or ancillary equipment.
   b. Dynamic stresses, such as shock and vibration.
   c. Gaseous and particulate contamination, such as exhaust or combustion emissions.
3. Combined natural and induced conditions. Frequently, the stresses affecting an item result from a combination of one or more factors from both classes. Such combinations may intensify the stress, or the combined factors may cancel each other out.

Provide for the Adequate Training of Personnel

The implementation of a reliability assurance program requires a training program at both the operational and supervisory levels. At the operator level, instruction should be given in the collection of failure and maintenance data, in the maintenance function (both preventive and unscheduled maintenance or repair of equipment), and in the control of operating conditions. This training can be accomplished by the use of lectures, demonstrations, films, posters, and reliability information bulletins.

At the supervisory level, in addition to the above, training should be given in the analysis of reported data, program planning, testing, and demonstration procedures.

The reliability of the measurement system depends, to a large extent, on the training of the operator. The completeness of the data, as measured by the proportion of valid data reported, is a function of both the reliability and maintainability of the instrumentation.

PROVIDE PREVENTIVE MAINTENANCE

In order to prevent or minimize the occurrence of wearout failure, the components of the system subject to wearout must be identified and a preventive maintenance scheduled implemented for them. This activity aids in improving the completeness of the data. This maintenance can be performed during nonoperational periods for noncontinuous monitoring equipment, resulting in no downtime. Replacement units should be employed in continuous monitoring systems in order to perform the maintenance while the system is performing its function. Scheduled downtime may also be employed.

CONSIDER MAINTAINABILITY AT THE TIME OF PURCHASE

Maintainability is the probability that the system will be returned to its operational state within a specified time after failure. For continuous monitoring instruments maintainability is an important consideration during procurement, and in some cases may be desirable to include in the purchase contract. Maintainability items to consider for inclusion at the time of procurement are:

1. Design factors:
   a. The number of moving parts.
   b. The number of highly-stressed parts.
   c. The number of heat-producing parts.

2. Ease of repair after failure has occurred.
3. Maintainability costs:
   a. Inventory of spare parts required.
   b. Amount of technician training required for repair.
   c. Factory service required.
   d. Service repair contract required.
   e. Estimated preventive maintenance required.

PROVIDE RECORDS OF FAILURE AND MAINTENANCE; ANALYZE AND USE TO INITIATE CORRECTIVE ACTION

Field reliability data should be collected in order to:

1. Provide information upon which to base reliability rate predictions.
2. Provide specific failure data for equipment improvement efforts.
3. Provide part of the information needed for corrective action recommendations.

References

Bazovsky, I. 1961. *Reliability Theory and Practice.* Englewood Cliffs, N.J.: Prentice-Hall.

Enrick, N. L. 1972. *Quality Control and Reliability, 6th Ed.* New York: The Industrial Press.

Haviland, R. P. 1964. *Engineering Reliability and Long Life Design.* Princeton, N.J.: D. Van Nostrand Co. Inc.

Juran, J. M. 1974. *Quality Control Handbook, 3rd Ed.* New York: McGraw-Hill Book Company.

Muench, J. O. 25–27 January, 1972. "A Complete Reliability Program." Paper read at Annual Reliability and Maintainability Symposium, Institute of Electrical and Electronic Engineers. San Francisco, CA.

1967. *MIL-STD 718B Reliability tests, Exponential Distribution.* Washington, D.C.: Department of Defense.

1975. *Quality Assurance Handbook for Air Pollution Measurement.* Research Triangle Park, N.C.: U.S. Environmental Protection Agency.

Part 2

HOW TO WRITE A LABORATORY QUALITY ASSURANCE MANUAL

SECTION 28   INTRODUCTION

All laboratories employ some sort of quality program or system, but some are more structured and identifiable than others.

When a laboratory, driven by regulatory, accreditation, or marketing pressures, decides that it must have a formal, written quality program described in a manual, the question of cost is immediately raised. As we have seen in the discussion of quality costs in Section 25, laboratory quality cost benefits are often not available or identifiable, due to the structure of the organization's accounting system. On the other hand, the efforts made to set up a new laboratory system involving new forms, the introduction of new procedures or changes to existing procedures, added techniques, and so forth are highly visible as expenses.

The purpose of this Part 2, then, is to disclose to quality practitioners in the laboratory a technique that will lead to the efficient development of a quality manual that describes a system that is not too cumbersome for the laboratory it supports and is not counterproductive because of excessive demands for paperwork and reports.

SECTION 29   ORGANIZING FOR PREPARATION OF THE MANUAL

Before dealing specifically with how to go about writing a laboratory quality manual, it would be wise here to firmly establish the purpose or purposes of the document. This is done in order to be able to keep constantly in mind, during the production of the manual, the reason for writing the documents, so that the contents will enable the organization to satisfy the requirements for which the manual is produced.

Basically, the quality manual is a document setting forth the laboratory's policies and operating procedures which will affect the quality of the laboratory "product," remembering that the precision and accuracy of analytical or test results are the measures of the laboratory's performance quality.

The Quality Control Coordinator is, normally, the individual assigned the task of producing the quality manual. Others may be assigned to assist him or her in gathering information and drafting portions of the manual. Additionally, other managers must be made aware of the program and their responsibility to provide procedures, forms, and information affecting laboratory quality.

The organizational steps necessary for producing the laboratory quality manual are:

- Establishment of quality objectives and policies (Sections 4 and 5).
- Collection and review of existing applicable procedures.
- Preparation of a flow chart.
- Identification of quality system requirements to be selected.
- Fitting of existing procedures to requirements.
- Identification of shortfalls.
- Establishment of priorities.
- Writing the manual.
- Reviewing and making necessary changes.

SECTION 30   ESTABLISHING OBJECTIVES AND PRIORITIES

Since the quality program should be developed to meet the requirements of accrediting bodies, government regulations, higher management, or marketing pressures, it is important that, in planning the laboratory quality program, the above considerations are incorporated into the system. It is essential at this stage that quality objectives and policies are established and clearly described. These descriptions must be in writing and should be endorsed by the signatures of top laboratory management (see Sections 4 and 5).

One way of obtaining the endorsement and support of management is for the Quality Control Coordinator to draft sets of recommended objectives and policies for submission to management for review. After discussion and revision, agreement is reached establishing the course of the laboratory's programmed activity, and management returns the endorsed objectives and policies to the Quality Control Coordinator.

Once the tenor of the laboratory's quality program has been established, planning can commence. Through either the formal publication of quality objectives and policies, or by a letter of promulgation, it should be made clear to all personnel in the organization that the quality manual is an expression of management's intent that its provisions are binding on all individuals and departments, sections, or branches of the laboratory, and is not a document meant just for the Quality Department. It is desirable, then, that as many individuals as possible, within the laboratory, participate in the preparation and drafting of the document.

SECTION 31   COLLECTION AND REVIEW OF EXISTING APPLICABLE PROCEDURES

Whether or not they consider present practices to be a part of a formal quality control system, all laboratories have existing procedures, formal and informal, written or unwritten, which guide their activities and affect the quality of laboratory output. Having established the quality objectives and policies for the laboratory, the next step is to establish an inventory of existing procedures which affect quality and govern the work, either formally or informally, within the organization or in its relationships with other organizations. Procedures are considered to be such things as: instructions on the use and distribution of forms and reports, Standing Operating Procedures, detailed explanations of policy statements, protocols for the conduct of inter-laboratory testing programs, or any other instruction, rule, or document that governs the conduct of laboratory operations and affects the quality of laboratory output.

In small laboratories, the Quality Control Coordinator may have most of the desired information at his fingertips. In larger organizations, it may be necessary to conduct interviews with other department heads and their subordinates, circulate questionnaires, hold meetings with outside agencies and spend a great deal of time and energy on preliminary investigation. In larger

organizations, where written policies and procedures have been in existence for some time, there very often exists a subculture of unwritten procedures which are those actually being followed. It is up to the investigator to ferret those out and bring them to the attention of higher management to decide if any have merit and should be brought into the formal system; or, conversely, if such unauthorized procedures are working to the detriment of the organization, take measures to eliminate their use. The end result of this activity should be a compilation of documents and forms which the Quality Control Coordinator will incorporate as a part of the final quality document.

It should be pointed out that a useful tool for keeping track of, and filing, the various documents that have been collected during the investigation is an "Every-Day File" fast sorter. After making up a tentative table of contents, the sorter can be used to file the collected documents in numerical order.

SECTION 32   PREPARATION OF A FLOW CHART

The next step in the preparation of the laboratory quality manual should be to draw up a detailed analytical or testing operations flow chart.

As an illustration, the operations of an industrial hygiene analytical laboratory will be used, showing two parallel flow paths that demonstrate what happens when a sample arrives at a laboratory, and moves through the laboratory to final disposition (Figure 5-1). We have then:

1. A flow chart of the path of the sample from receipt through the analytical or test cycle, until the report of results is rendered and disposition of the sample is made.
2. A flow chart of the documentation accompanying the sample, as well as that generated through the analytical or test cycle. The chart begins with the Sample Test Request, and follows through the analytical or test cycle, preparation of the report, and on to distribution of the copies.

Preparing the flow chart also provides the opportunity to scrutinize and identify the details of laboratory operations that affect the quality of laboratory output and provide the ground work for the next step in the preparation of the manual, identification of the Quality System Elements to be selected.

In the smaller laboratory, the person designated to prepare the manual may have enough familiarity with the detailed operations of the organization so that he or she can prepare the flow chart without having to seek assistance from outside sources. In larger laboratories, assistance in the preparation of the flow chart may be needed from each different department, section, or

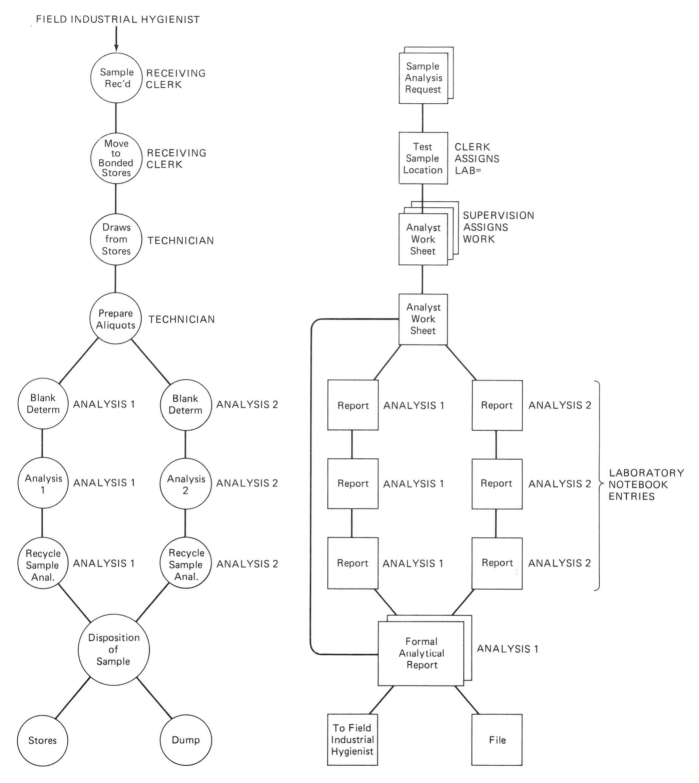

Figure 5-1

branch involved, after which the results are combined to establish the flow of work through the laboratory. Given this information, the Quality Control Coordinator or other person preparing the manual should be able to identify those activities which relate to the quality control function. It is during this period that laboratory procedural subcultures may be discovered. It then behooves the writer to decide upon the merit of such unprogrammed activity and take action to adopt or to eliminate it. The next step is to critically review the flow chart to make sure, first, that it is a fair representation of the procedural system in the laboratory, and

33

second, if it is, to determine whether there is any need for improvement to increase efficiency by eliminating or combining steps to streamline operations.

SECTION 33 IDENTIFICATION OF PROGRAM REQUIREMENTS

By this time, the outline of the content of the manual should be taking shape in the mind of the person preparing the manual, and he or she should be in a position to prepare a tentative table of contents. Using the information derived from taking an inventory of existing procedures, taking into consideration requirements generated by the establishment of quality goals and policies and reviewing the requirements of regulatory or accrediting bodies, a list of elements that will comprise the laboratory's quality system can be prepared.

SECTION 34 IDENTIFICATION OF SHORTFALLS AND THE ASSIGNMENT OF PRIORITIES

Once the list of elements is prepared, and all are in agreement that this is acceptable as the way for the laboratory to conduct its quality business, the next step is to identify the procedures that are already in place within the organization and then, further, to identify those which are not in existence and must be developed. It is here that the Every-day File becomes useful, since it can be used to accumulate the various written procedures and forms applicable to each element and file them by subject matter. The elements that need to be developed will have blank pockets, and thus can be easily identified. Once having established what new procedures need to be developed in order to complete the manual in the manner desired, it will be necessary to establish a priority for accomplishing these tasks. Obviously, not all new procedures can be developed and put into place simultaneously, so the person preparing the manual must decide the sequence in which the new material is to be developed. Once these new procedures have been developed, the writer is ready for the final step.

SECTION 35 WRITING THE MANUAL

Presented here are tips and suggestions made to assist the writer in preparing a more acceptable document which will meet the requirements of most regulatory and accrediting organizations.

1. Start with a Title Page (Section 1), followed by a Table of Contents. A Letter of Promulgation signed by the highest available authority is not essential, but is a definite asset if the laboratory management will actively and demonstrably pursue its commitment to quality. If the Letter of Promulgation is merely lip service to a requirement grudgingly accepted, then the letter is worse than useless, and should be omitted.

2. There is no particular order in which the Table of Contents should be listed, or in the order of appearance of the various elements described in the manual. It seems logical, however, that the Goals, Policies, and Organization Sections of the manual should appear early, and in that order.

3. The pages of each section of the manual should be numbered sequentially, beginning with "1" with each new section. This makes it easy to insert new pages, should it become necessary at a later date, without having to renumber all subsequent pages of the document.

4. The paragraph numbering system should be uniform throughout the manual. The system may be numeric or alphanumeric but should be the same in each section of the document.

5. The format of each section, once established, should be kept consistent throughout the document.

6. Blank copies of all forms, tags, labels, stickers, reports, and so on, whenever mentioned in the text of the manual, should be included, either directly following the page on which referenced, or, in order, at the end of each section. For easy reference, this is preferable to incorporating all forms at the end of the entire document. The latter is not unacceptable, but in the case of a large organization, having many formalized procedures, this practice becomes inconvenient. Each form should be numbered and referenced back to the point of mention in the text.

Part 3

XYZ LABORATORY QUALITY ASSURANCE MANUAL

INTRODUCTION

This part contains a representative example of a Laboratory Quality Assurance Manual. It is intended to be used as a model to be copied, amended, supplemented or extracted from in order to tailor a document that will accurately describe a particular laboratory's operations. It may also be used as a model against which a laboratory's existing manual may be compared to determine how the latter document compares with current doctrine.

The sample manual incorporated 27 elements of a typical quality system. The elements are those prescribed by ANSI/ASQC Q2-19XX. However, many laboratories will find that they do not use nor will they choose to incorporate all of the elements described. Selection of the individual organization's components for its particular way of conducting laboratory operations depends on regulatory or accrediting requirements, management decisions, and marketing pressures.

This part should be used in conjunction with the descriptive material in Part 1. The organizational structures, job descriptions, procedural details, and forms, should, of course, be amended and designed to reflect the user's operations.

QUALITY ASSURANCE MANUAL

XYZ LABORATORIES, INC.

1111 Main Street

Anytown, MA 45222

Prepared by: _____

Approved: _____

Date: _____

Copy No.: _____

SECTION   TITLE

39

This manual is issued to describe the quality assurance system employed at XYZ Laboratories, Inc. in compliance with the intent of the general quality system requirements of the cognizant accrediting organization. The policy of XYZ Laboratories, Inc. is to apply the system to all testing and analytical activities undertaken on behalf of the customers or an accrediting organization.

The manual provides personnel and customers of XYZ Laboratories, Inc. with a description of company policy for maintaining an effective and economical quality assurance system planned and developed in conjunction with other management planning functions.

Written procedures for implementing the policies described herein are amplified by the several sections comprising this manual. These procedures are binding on all personnel of the laboratory and shall be adhered to implicitly.

The Quality Assurance Program described in this Quality Assurance Manual has the absolute and unqualified support of XYZ Laboratories, Inc. Management.

Our established goal–deliverance of highest quality service at a fair price is the same today as when the laboratory was founded in 1932.

Our testing and analytical services–their precision and accuracy, the care with which they are conducted, and their customer acceptance–are the means by which XYZ Laboratories, Inc. has gained an enviable reputation and has become a leader in the industry. Quality leadership is our number one priority and every member of the laboratory staff shares the responsibility of maintaining our present status.

At XYZ Laboratories, Inc., quality is the responsibility of every employee. Meeting this commitment will result in the continued satisfaction of our customers and an improved quality of life for our employees.

The provisions of this manual, describing the XYZ Laboratories, Inc. Quality System, are binding on those individuals given the responsibilities outlined herein.

I will expect everyone concerned to use this manual as a guide to the continued maintenance and improvement of the quality of our laboratory services.

Sincerely,

3.1 <u>Purpose</u>

The purpose of this section is to delineate the Quality Objectives of XYZ Laboratories, Inc.

3.2 <u>Scope</u>

The objective of the Laboratory Quality Assurance Program is to assure the accuracy and precision, as well as the reliability of laboratory results produced for our customers, or at the request of regulatory or accrediting bodies. Management, administrative, statistical, investigative, preventive, and corrective techniques will be employed to maximize reliability of the data.

3.3 Specific objectives are:

3.3.1 To develop and put into service rugged methods capable of meeting the user's needs for precision, accuracy, sensitivity, and specificity.

3.3.2 To ensure that all staff members receive training in basic quality technology, in sufficient depth to enable them to carry out the provisions of this manual. In addition, the Quality Control Coordinator shall receive sufficient additional training to enable him or her to receive American Society for Quality Control Certification within the prescribed period.

3.3.3 To establish the level of quality of the laboratory's routine performance as a baseline against which to measure the effectiveness of quality improvement efforts.

3.3.4 To make any changes in routine methodology found necessary to make it compatible with performance needs as established in Par. 3.3.1 above.

3.3.5 To monitor the routine operational performance of the laboratory through participation in appropriate inter-laboratory testing programs and to provide for corrective actions as necessary.

3.3.6 To improve and validate laboratory methodologies by participation in method validation studies.

4.1 <u>Purpose</u>

This section lists policies to be implemented by the laboratory in order to achieve the objectives set forth in Section 3 and in the furtherance of the overall quality control program.

4.2 <u>Scope</u>

This section sets forth only the outlines of management's policies with regard to Quality Assurance. Details for carrying out these policies appear in later sections of the manual.

4.3 Laboratory quality policies include:

4.3.1 Quality activities shall emphasize the prevention of quality problems rather than detection and correction of problems after they occur.

4.3.2 Quality Cost figures shall be computed quarterly and reported to management.

4.3.3 All Employees engaged in making decisions affecting the quality of laboratory output shall undergo training programs designed to be commensurate with their positions, duties, and responsibilities.

4.3.4 The laboratory shall use published analytical and test methods wherever available.

4.3.5 The laboratory shall retain copies of all test and analytical reports in a manner and for a period specified by regulatory or accrediting bodies.

4.3.6 The laboratory shall have a comprehensive calibration

program involving all instrumentation used for making determinations, the results of which are reported.

4.3.7   The laboratory shall use appropriate, fresh reagents and chemicals, certified when necessary, and appropriate calibrated glassware.

4.3.8   The laboratory shall establish and maintain a total intralaboratory quality control system to assure continued precision and accuracy of laboratory results.

4.3.9   The laboratory shall participate in an interlaboratory testing program as prescribed by the cognizant accrediting organization.

5.1 <u>Purpose</u>

This section describes the organization of XYZ Laboratories, Inc. and provides a copy of the Job Description of the Quality Control Coordinator.

5.2 <u>Scope</u>

This section is designed to identify the position of Quality Assurance Management within the overall structure of the laboratory organization. Other position descriptions are found in personnel department files.

5.3 The management of a Quality Control Program as described in this manual requires the services of a Quality Control Coordinator within the laboratory to carry out the monitoring, record-keeping, statistical techniques, calibration, and other functions required by the system. The Quality Control Coordinator of XYZ Laboratories, Inc. reports to the Laboratory Director, who is also the President of the corporation.

5.4 The Organization Chart of XYZ Laboratories, Inc. illustrating the placement of the quality function within the organization follows as Figure 5-1.

5.5 The Job Description of The Quality Control Coordinator appears as Figure 5-2.

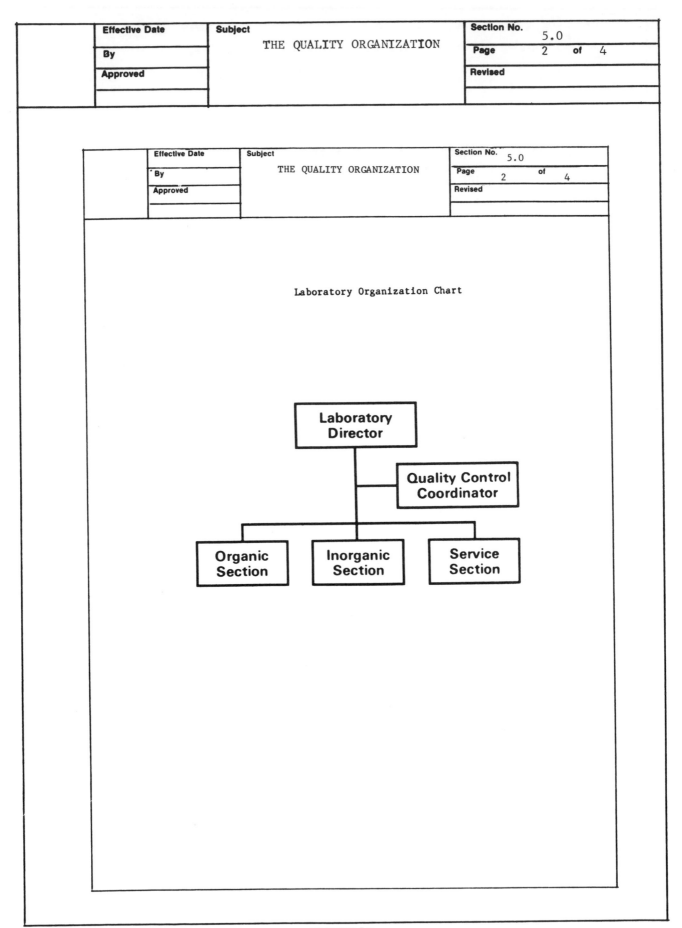

Laboratory Organization Chart

FIGURE 5-1

<u>JOB DESCRIPTION</u> (Figure 5-2)

<u>TITLE:</u> Quality Control Coordinator

1. Basic Function

   The Quality Control Coordinator is responsible for the conduct of the XYZ Laboratories, Inc. Quality Control program and for taking or recommending measures to ensure the fulfillment of the quality objectives of management and the carrying out of Quality Policies in the most efficient and economical manner commensurate with ensuring continuing accuracy and precision of data produced.

2. Responsibilities and Authority:

   2.1 Develops and carries out quality control programs, including statistical procedures and techniques, which will enable the laboratory to meet desired quality standards at minimum cost; and advises and assists management in the installation, staffing, and supervision of such programs.

   2.2 Monitors quality control activities of the laboratory to determine conformance with authorized policies and procedures and with sound practice; and makes appropriate recommendations for correction and improvement as may be necessary.

   2.3 Seeks out and evaluates new ideas and current developments in the field of quality control and recommends means for their application wherever advisable.

Figure 5-2

2.4   Reviews new technology, methods, and equipment and advises management as to such use, with respect to quality aspects.

2.5   Advises the Purchasing Component with regard to the quality of purchased equipment, materials, reagents, and chemicals.

2.6   Recommends packaging materials and procedures as well as necessary changes thereto.

2.7   Performs such other related duties as may be assigned.

6.1 Purpose

The purpose of this section is to define the tasks and responsibilities relating to the preparation, distribution, review, and maintenance of the Quality Manual.

6.2 Scope

This section deals primarily with those manuals which are issued under controlled conditions.

6.3 Issues, Distribution, and Maintenance of the Manual

The Quality Control Coordinator bears the primary responsibility for the preparation, issue, review, and upkeep of the Laboratory Quality Manual.

After the preparation of the manual, the Quality Control Coordinator is responsible for the initial distribution of controlled copies of the manual. Controlled copies are serially numbered, and a distribution list is kept showing to whom each copy has been issued.

The purpose of this control is to make sure that changes are distributed to recipients of the manual when necessary and that copies are retrieved when personnel changes require that the manual no longer be in the hands of the affected individual.

Uncontrolled copies of the manual may be distributed, from time-to-time to individuals or organizations outside the laboratory. These copies will not be numbered or logged and will not receive changes as they occur. Uncontrolled copies will be so marked.

The Quality Control Coordinator is responsible for the timely, periodic review of the content of the manual to ensure that its requirements reflect current operating conditions and meet with needs. This will normally be done immediately following internal quality system audits or audits, assessments, or site visits by an outside accrediting organization.

This manual is a numbered, controlled document, unless otherwise marked. Revisions, additions, or deletions occurring as a result of periodic review or other authorized changes will be controlled through the issue of revisions to individuals or laboratory components listed on the master distribution list.

6.4 <u>Preparation</u>

6.4.1 The Quality Assurance Manual will be published using the standard format on which this Section appears. For details on the use of this format, see Section 27.0

6.4.2 This manual will use the alpha-numeric system of paragraphing as illustrated by Par. 11.2.2 ff. and example:

      1.1

      1.2

         1.2.1

         1.2.2

            a.

7.1 <u>Purpose</u>

The purpose of this section is to establish a method for developing and maintaining detailed plans which will provide for the quality aspects of producing and delivering precise, accurate tests or analytical results for our customers.

7.2 <u>Scope</u>

Quality Assurance Planning takes place in two phases—the initial phase takes place when the laboratory quality system is being developed and installed. The planned Quality System elements, gathered in an authoritative collection of written procedures and their accompanying forms constitute the Quality Assurance Manual.

It is this first phase which is discussed in this section. The second phase is that planning which takes place during the implementation and conduct of the quality program on a continuing basis. This phase has been discussed in the preceding Section 6.0., involving the handling of changes in procedures brought about by changes in technology, personnel, regulatory or accrediting requirements, management decisions, etc., which in turn result in changes to the Laboratory Quality Assurance Manual.

7.3 Approach to Planning

The act of planning is thinking out in advance the sequence of actions to accomplish a proposed course of action in doing work to accomplish certain objectives. In order that the planner may communicate his plan to the person or persons expected to execute it, the plan is written out together with necessary criteria, diagrams, tables, etc.

Planning for laboratory quality assurance must fundamentally be geared for delivering acceptable quality data at an acceptable cost. These objectives are realized only by carefully planning many individual elements which relate properly to each other and that make up the Quality Program Requirements set forth in this manual.

7.4 Development of a Quality Plan

The first step in the quality planning sequence is to determine which quality assurance elements should be included as a part of a quality assurance plan. (QA Plan). This is done by taking into consideration:

1. Management direction.

2. Regulatory or accrediting requirements.

3. Applicable consensus quality standards.

4. Budgetary and cost considerations.

5. Industry practices.

With these considerations in mind, a preliminary list should be prepared, tentatively setting out the elements chosen to comprise the laboratory's quality system.

Having chosen a tentative list of quality elements, the next step is to inventory existing procedures which will be absorbed readily into the quality system. Following this, shortfalls should be identified and procedures developed to fill out the quality plan.

The culmination of the initial quality planning should be a document which includes the essential information, directives, and documents or forms which will make up the laboratory Quality Assurance Manual.

The XYZ Laboratories, Inc. Quality Assurance Manual is designed to serve several functions:

1. Most importantly, it is the result of a planning effort to design a system which will ensure that our laboratory's results will be of the highest precision and accuracy.

2. It represents an historical record which documents the procedures and subsystems established to control laboratory performance quality.

3. It provides management with a document which can be used to assess whether quality control and quality assurance activities described are being implemented. In other words, it provides a check list against which the quality program can be audited.

4. It will be used as a training document for the indoctrination of new employees.

5. It provides the description of The XYZ Laboratories, Inc. Quality System which can be used as evidence in defense in the case of litigation involving the validity of test or analytical reports.

6. It will serve to meet requirements of regulatory or accrediting organizations.

7. It will be used as a sales tool to support the laboratory's reputation for the production of high quality work.

8. It describes to all concerned the elements of the XYZ Laboratories, Inc. quality system.

8.1 Purpose

This section describes the procedures for including technical and quality requirements in purchase orders and for marking, receiving, storing and issuing reagents, chemicals, and testing supplies and materials.

8.2 Scope

This section deals only with supplies and materials used in testing and analytical work. It is not concerned with test or analytical samples or testing equipment.

8.3 A vendor of testing or analytical supplies and materials is regarded as a resource to and an extension of the laboratory organization. The standards for quality, therefore, imposed on vendors are the same as those self-imposed on the laboratory.

8.4 The Purchase Order instructs vendors to mark packing slips and containers of reagents, chemical and testing supplies with the following information, when applicable:

Name of material

Vendor's name and address

Vendor's lot number

Quantity

Material specification number and date

Purchase order number

This assures that the material is properly identified and that the vendor is using the latest specification.

8.5   Copies of all purchase orders for all reagents, and testing and analytical supplies are sent to the Laboratory Quality Control Coordinator where they are reviewed to assure the latest requirements are correctly specified.

8.6   Purchase orders, receiving documents and accompanying certifications are used as part of the receiving control procedure and show information necessary to identify the material being received.

8.7   Control of Incoming Materials

   8.7.1   The Laboratory Stores Clerk segregates incoming reagents, analytical and testing materials, and prepares the receiving report (see Figure 8-1) as follows: (1) the date of receipt, (2) The name of the shipper or vendor, (3) the point of origin, (4) the shipping routing or method, (5) the shipping charges, (6) method of payment of shipping charges, (7) the freight bill number, (8) the Purchase Order number, (9) the laboratory component originally ordering the item, (10) the number of items listed on the packing slip, verified by count, (11) the description of the materials, (12) the condition of the items received, i.e., whether they are accepted or not, (13) any pertinent remarks, such as the reason for rejection, (14) the signature of the Receiving or Stores Clerk.

8.7.2 Testing materials received in large quantities, such as gas detector tubes and sampling filters may be subjected to incoming inspection using ANSI/ASQC Z1.4 sampling plans or other approved sampling plans. Written Inspection Instructions will be used during Incoming Inspection. The Quality Control Coordinator will prepare the Classification of Defects/Inspection Instructions form (shown in Figure 8-2) as follows: (1) the item description, (2) the part or model number, (3) the lot size, (4) the detailed inspection or test instructions, making reference to test specification documents, if necessary, (5) indicate gages, instruments, or equipment required for the inspection or test.

8.7.3 The Stores Clerk will check the package marking and the Packing Slip against the Quality Control copy of the Purchase Order. If a discrepancy is found that could affect analytical or testing accuracy or precision, the material is discarded, Purchasing is notified, and the disposition is noted on the Receiving and Stores Log. (Shown as Figure 8-3.) The Stores Clerk is responsible for posting the Receiving and Stores Log as follows: (1) each incoming shipment is assigned a serial number which is entered on the Log, (2) the item description is entered, (3) the Purchase Order number for the shipment is entered, (4) the Vendor's name is entered, (5) the date of receipt is entered, (6) the disposition of the shipment, i.e., whether or not it was accepted or rejected is entered. If the material is accepted, the shipment is logged in and placed in Stores with appropriate log entries made as above.

8.7.4 Materials are placed in Stores in ''last-in, last out'' order. Materials having a finite shelf-life are plainly marked with an expiration date.

8.7.5 Materials are issued, on request, on a ''first-in, first-out'' basis. Before issue, containers are checked to make sure that expiration dates are not exceeded.

8.7.6 When it is necessary to conduct an audit of a vendor's quality system, the Quality System Survey Evaluation Check List for vendor audits will be used (Figure 8-4). The first page of the Check List will be filled out as follows: (1) Enter the name of the company being audited. (2) Enter the auditee's address. (3) Enter the auditee's phone number. (4) List the general type of products or services furnished by the company. (5) Indicate the total number of employees at this location. (6) Indicate the number of employees assigned to the quality function. (7) List the names of the individuals with whom the auditor or audit team was in contact during the audit, and list the products considered during the course of the audit. (8) Indicate the name or names of those conducting the survey, or audit. Enter the date or dates of the audit. (9) Enter the survey results including the percentage scored.

XYZ LABORATORIES, INC.

RECEIVING REPORT

Date _____(1)_____ 19 _____

Received From _____(2)_____

Shipped From. _____(3)_____

Shipped Via _____(4)____

SHIPPING CHARGE (5)	PREPAID (6)	COLLECT	FREIGHT BILL NO (7)	

PURCHASE ORDER NO (8)		FOR DEPT (9)	

QUANTITY	DESCRIPTION OF MATERIALS	CONDITION
(10)	(11)	(12)

Remarks: _____(13)_____

RECEIVING
CLERK

FIGURE 8-1

XYZ LABORATORIES, INC.
QUALITY ASSURANCE
CLASSIFICATION OF DEFECTS
INSPECTION INSTRUCTIONS

DESCRIPTION (1) _____

PART NUMBER (2) _____ LOT SIZE (3) _____

SAMPLING PROCEDURE: ANSI / ASQC / Z 1.4 Level II, unless otherwise noted.

CRITICAL
100% (no defects allowed)

INSPECTION PROCEDURE	METHOD
1: _____(4)_____	_____(5)_____
2. _____	
3. _____	
4. _____	
5. _____	
6. _____	

MAJOR "A"
AQL 1.0 %

101. _____	
102. _____	
103. _____	
104. _____	
105. _____	
106. _____	

MAJOR "B"
AQL 2.5 %

201. _____	
202. _____	
203. _____	
204. _____	
205. _____	

MINOR
AQL 4 %

301. _____	
302. _____	
303. _____	
304. _____	
305. _____	

FIGURE 8-2

TESTING AND ANALYTICAL SUPPLIES

RECEIVING AND STORES LOG

Log No.	Identification	P.O. No.	Vendor	Date	Disposition Acc. - Rej.
(1)	(2)	(3)	(4)	(5)	(6)

FIGURE 8-3

XYZ LABORATORIES, INC.

QUALITY SYSTEM SURVEY EVALUATION CHECK LIST

COMPANY _____(1)_____

ADDRESS _____(2)_____ PHONE ____(3)_____

GENERAL PRODUCTS __(4)_____

TOTAL NUMBER OF EMPLOYEES ___(5)_____ .QC ___(6)_____

PERSONNEL CONTACTED AND PRODUCTS CONSIDERED (7)

SURVEY BY AND DATE   (8)

_____   SURVEY RESULTS ____(9)_ OF _____=_____ %

_____

_____

_____

NAME — TITLE — DATE

FIGURE 8-4

INDEX

SURVEY RESULTS

INSTRUCTIONS

An applicant's compliance to each question shall be indicated by placing 0, 1, 2, 3, or NA on the line opposite each question.

0. Required but not being done.
1. Below acceptance standards.
2. Adequate with minor departures from good practices.
3. Adequate in all respects.
NA. (Not Applicable) when it is obvious that a question would be inappropriate, due to type of product, lack of facilities, special processes or other valid reasons.

At the conclusion of the survey, each category of questions should be analyzed and summarized for acceptability. Space is provided in the index for entering these results. Remarks or comments by the survey team will be placed on the blank page opposite the question.

The following method shall be used in summarizing each category: Total the survey points for all applicable questions. Divide this total by the total applicable questions. Multiplied by the weighting factor of 3, this computation will result in a percent acceptability figure.

$$\frac{\text{Total points for applicable questions rated}}{\text{Number of applicable questions X 3}} = \text{Percent Acceptability}$$

Example:
For category 1, we have a total of 10 applicable questions and a weighting factor of 3 per question (10 X 3)

Therefore $\frac{28}{10 \text{ X } 3} = 93\%$ (Percent acceptability for category)

CATEGORY 1
QUALITY CONTROL ORGANIZATION

1. Company management reflects a positive attitude toward Quality Control. _____
2. Direct responsibility for Quality Control has been formally established. _____
3. The functional relationship between Quality Control and other departments has been clearly delineated. _____
4. Line of authority within the Quality Control organization has been clearly delineated. _____
5. The Quality Control System has been designed to promote defect prevention. _____
6. The Quality Control System is adequately documented. _____
7. The Quality Control Department issues periodic reports to management which reflect quality levels, rework, scrap, work realization, quality costs, etc. _____
8. The documented Quality Control System carries management approval. _____
9. The Quality Control Department adequately provides for the use and control of inspection stamps. _____
10. The applicant provides for employee indoctrination and training in Quality Control methods. _____
11. Quality Control System assures that Engineering drawings and specification requirements, procured materials and manufactured items satisfy all customer contractual requirements. _____

Total Applicable Points _____
Total Survey Points _____
Percentage of Acceptability _____

- 1 -

CATEGORY 2
CONTROL OF PROCURED SUPPLIES

1. OPERATION AND DOCUMENTATION
   a. Inspection is preformed by Quality Control. _____
   b. Inspection instruction/procedures are available to inspection personnel. _____
   c. Inspection personnel perform inspection operations in accordance with current instructions and procedures. _____
   d. Applicant has provided a system to maintain instruction/procedures current. _____

2. VENDOR PROGRAM
   a. Applicant's Quality Control reviews environmental and life requirements and assures that Qualified Products are being procured. _____
   b. Applicant has an established program for processing drawing changes to his vendors and subcontractors. _____
   c. Receiving Inspection records reflect quality history of applicant's vendors and subcontractors. _____
   d. Vendor quality performance data reports are used by other departments in making procurement decisions. _____
   e. Applicant has program for Quality Control approval of vendors and subcontractors. _____
   f. Applicant performs source surveillance at subcontractor's plant when applicable. _____
   g. Applicant requires his vendors and subcontractors to have a system for quality control. _____
   h. Applicant uses inspection/test results to contribute to vendors evaluation. _____

3. INSPECTION CRITERIA
   Purchase Orders, Drawings, Engineering Orders, specifications, and vendor's catalogs are available to Receiving Inspection. _____

4. INSTRUMENTATION
   a. Inspection gages and test equipment are adequate to perform required inspection. _____
   b. Tools, fixtures, and inspection equipment are identified, stored and issued under controlled conditions. _____

5. SAMPLING INSPECTION (When Applicable)
   a. Sampling Inspection is performed to MIL-STD-105, MIL-STD-414, or other approved plan. _____
   b. Inspection personnel are provided with instructions covering Sampling Inspection. _____

6. CONTRACTED SPECIAL PROCESSES
   a. Applicant has established certified and/or approved source for special process services. _____
   b. Applicant's Quality Control performs surveillance over special processing. _____

7. STOCK CONTROL
   a. Applicant uses a positive means of identification on all stock. _____
   b. Applicant requires and receives chemical/physical analysis or test specifications when applicable _____
   c. Applicant provides adequate control area for customer furnished material when required. _____
   d. Applicant has an acceptable system for "age control" for items where acceptability is limited by maximum age. _____

   Total Applicable Points _____
   Total Survey Points _____
   Percentage of Acceptability _____

- 2 -

CATEGORY 3
IN-PROCESS INSPECTION

1. **OPERATION AND DOCUMENTATION**
   a. In-process inspection is performed by Quality Control. _____
   b. Inspection instructions/procedures are available to Inspection personnel. _____
   c. Inspection personnel perform inspection operations in accordance with current instruction and procedures. _____
   d. Applicant has provided a system for First Article Inspection and re-inspection after changes in manufacturing process. _____
   e. Applicant procedures provide for inspection control of in-process material. _____
2. **INSPECTION CRITERIA**
   a. Illegible or obsolete drawings are not in use by inspection. _____
   b. Current Drawings, Engineering Orders, and specifications are available at Inspection area. _____
3. **INSTRUMENTATION**
   a. Adequate in-process inspection facilities are available. _____
   b. Inspection gages and test equipment are adequate to perform required inspection. _____
   c. Adequate surveillance and maintenance of inspection equipment when in use, transported or stored is maintained. _____
4. **STATISTICAL QUALITY CONTROL** (When Applicable)
   a. Sampling inspection is preformed to MIL-STD-105. MIL-STD-414, or other approved plan. _____
   b. Inspection personnel are provided with instructions covering sampling inspection. _____
   c. Control charts or other in-process statistical quality control methods are used. _____

Total Applicable Points _____
Total Survey Points _____
Percentage of Acceptance _____

CATEGORY 4
FINAL INSPECTION

1. **OPERATION AND DOCUMENTATION**
   a. Final inspection is performed by Quailty Control. _____
   b. Current Inspection instruction/procedures are available to Inspection personnel. _____
   c. Inspection personnel perform inspection operations in accordance with current instructions and procedures. _____
   d. Completed supplies are inspected as necessary to assure that contract requirements have been met. _____
   e. Environmental and life tests are periodically performed to assure that manufacturing and process degradation has not significantly effected design integrity. _____
2. **INSPECTION CRITERIA**
   Current Drawings, Engineering Orders, specifications and/or customer requirements are available at final inspection. _____
3. **INSTRUMENTATION**
   a. Inspection gages and test equipment are adequate to perform required inspection. _____
   b. Tools, fixtures, and inspection equipment are identified, stored, and issued under controlled conditions. _____
   c. Adequate final inspection facilities are available. _____
4. **STATISTICAL QUALITY CONTROL** (When Applicable)
   a. Sampling Inspection is performed to MIL-STD-105, MIL-STD-414, or other approved plan. _____

- 3 -

CATEGORY 4
FINAL INSPECTION
(Cont.)

    b.   Inspection personnel are provided with instructions covering sampling inspection. _____

    c.   Control charts or other statistical quality control methods are used. _____

5.   **MATERIAL HANDLING**

    Provisions are made to prevent unauthorized use of uninspected and/or conforming
material. _____

                            Total Applicable Points _____

                            Total Survey Points _____

                            Percentage of Acceptability _____

CATEGORY 5
SHIPPING INSPECTION

1.   **OPERATION AND DOCUMENTATION**

    a.   Shipping Inspection is under the surveillance of Quality Control. _____

    b.   Inspection operations are performed in accordance with current instructions
and procedures. _____

2.   **INSPECTION CRITERIA**

    a.   Inspectors have packaging requirements. _____

    b.   Packaging tests are performed/witnessed as required by applicable specifications. _____

    c.   Certified packaging materials are used where required. _____

                            Total Applicable Points _____

                            Total Survey Points _____

                            Percentage of Acceptability _____

CATEGORY 6
CALIBRATION OF MECHANICAL MEASURING EQUIPMENT

1.   Quality Control procedures insure that inspection gages, measuring devices, and test
equipment are periodically inspected and recalibrated at established intervals. _____

2.   Production tooling which is in use as a medium of inspection is periodically inspected
by Quality Control at intervals that assure the maintenance of required accuracy. _____

3.   Applicant has system for modification of inspection/test equipment to latest engineering changes. _____

4.   Applicant does maintain, in a suitable environment, working standards of required
accuracy that are periodically calibrated to primary standards traceable to the National
Bureau of Standards. _____

5.   Personally owned tools and gages show evidence of periodic recalibration. _____

6.   Specialized test equipment used for acceptance purposes is proved satisfactory by
Quality Control before release for use. _____

7.   New or reworked test/inspection equipment is inspected and calibrated prior to use. _____

8.   Applicant has written instruction/procedures for "gage and test calibration." _____

9.   Objective evidence includes calibration date and date next calibration must be performed on measuring instrumentation. _____

10. Records are maintained for periodic recalibration of inspection gages and test equipment. _____

11. Record system provides for automatic recall of inspection gages and test equipment. _____

                            Total Applicable Points _____

                            Total Survey Points _____

                            Percentage of Acceptability _____

- 4 -

CATEGORY 7
CALIBRATION OF ALL OTHER TEST EQUIPMENT

1. Quality Control procedures insure that inspection equipment is periodically inspected and recalibrated at established intervals. _____
2. Production tooling which is in use as a medium of inspection is periodically inspected by Quality Control at intervals that assure the maintenance of required accuracy. _____
3. Applicant has system for modification of inspection/test equipment to latest engineering changes. _____
4. Applicant does maintain, in a suitable environment, working standards of required accuracy that are periodically calibrated to primary standards traceable to the National Bureau of Standards. _____
5. Personally owned tools show evidence of periodic recalibration.
6. Specialized test equipment used for acceptance purposes is "proved" by Quality Control before release for use. _____
7. New or reworked test/inspection equipment is inspected and calibrated prior to use. _____
8. Applicant has written instruction/procedures for all Measuring Equipment Calibration. _____
9. Objective evidence includes calibration date and date next calibration must be performed on measuring instrument. _____
10. Records are maintained for periodic recalibration of all inspection measuring equipment. _____
11. Record system provides for automatic recall of all inspection measuring equipment. _____

> Total Applicant Points _____
> Total Survey Points _____
> Percentage of Acceptability _____

CATEGORY 8
DRAWING AND CHANGE CONTROL

1. The direct and specific responsibility to verify that changes are incorporated at effective points is clearly defined and formally established. _____
2. Applicant has written procedures describing drawing change control. _____
3. A control system exists for the issue and return of each drawing. _____
4. Drawing and change control system prevents the use of illegible or obsolete drawings by Inspection. _____

> Total Applicable Points _____
> Total Survey Points _____
> Percentage of Acceptability _____

CATEGORY 9
ENVIRONMENTAL TEST FACILITIES
(When Applicable)

1. Applicant has suitable environmental test equipment to perform full range of required tests on product under consideration, or uses recognized test laboratories. _____
2. Applicant maintains records that show results of tests performed and failure rates. _____
3. Applicant has established program of calibration and maintenance of instrumentation used in environmental laboratory. _____

> Total Applicable Points _____
> Total Survey Points _____
> Percentage of Acceptability _____

- 5 -

CATEGORY 10
NON-CONFORMING MATERIAL
(Materials Review)

1. Applicant's procedures reflects control of authority to repair non-conforming supplies. _____
2. Applicant has a system for the diversion of non-conforming supplies from normal production flow.
3. Applicant has method of recurrence control to prevent repetitive discrepancies. _____
4. Corrective action requests on non-conforming materials are initiated promptly. _____
5. Records provide for follow-up on all corrective action requests. _____
6. Disposition of non-conforming material does not establish criteria for like dispositions. _____
7. Reports on non-conforming materials are issued to management for action. _____

Total Applicable Points _____
Total Survey Points _____
Percentage of Acceptability _____

CATEGORY 11
HOUSEKEEPING, STORAGE & HANDLING

1. Materials, supplies, and work in process are arranged in a neat workmanlike manner. _____
2. Work and storage areas are clean and free from dirt, rejected materials, and other materials which could contaminate or damage acceptable materials. _____
3. In the case of extended or indefinite storage, equipment is given proper preservative treatment. _____
4. Facilities for storage of materials or equipment assures prevention of deterioration and contamination. _____
5. During storage, material is packaged in a manner to prevent deterioration and damage. _____
6. Delicate instruments are handled in a manner that will not jeopardize the reliability of the design characteristics. _____
7. Transfer procedures are adequate to prevent comingling, contamination, and loss. _____
8. Tags and labels are properly used to indicate the identity, condition, and status of the equipment. _____
9. Components, assemblies, and materials are handled so as to prevent damage and deterioration. _____
10. Provisions are made to prevent unauthorized use of uninspected materials. _____

Total Applicable Points _____
Total Survey Points _____
Percentage of Acceptability _____

- 6 -

9.1 Purpose

The purpose of this section is to describe the duties and responsibilities of the Quality Control Coordinator with respect to shipping, packaging, handling and storage of samples.

9.2 Scope

This section provides guidance in making decisions pertinent to the validity and acceptability of samples submitted for testing or analysis. While it is particularly pertinent to samples submitted to the laboratory for chemical analysis, its principles apply broadly to all types of samples, the goal being the preservation of the integrity of the sample. It is applicable to all in-house and contract laboratory activities dealing with the handling of samples.

9.3 Physical Condition of the Sample Container

Physical damage to the sample container may be the fault of the carrier due to abusive handling or may be the fault of the sender because of faulty packaging. If damage to the container is evident, the package will be carefully opened and its contents inspected. In the event of damage to the sample because of damage to the shipping container, the sender will be notified and the invalid sample discarded. Where contract laboratory samples are involved, it will be the responsibility of the contractor to notify the Quality Control Coordinator concerning any suspect samples. He will then contact the sender and make any necessary decision regarding sample disposition.

9.4 Sample Integrity

Sample integrity refers to the cumulative end result of those factors which contribute to the overall validity of a sample. Sample integrity is promoted and preserved by adhering to adequate custodial, handling, and identification procedures by those individuals collecting the samples, up to the point of receipt of samples by the laboratory. When the samples are received for testing or analysis they are checked for:

1. Physical damage to samples because of inadequate packing and protection.

2. Loss of samples because of inadequate or improper sealing. This includes leakage of liquids from vials, loss of particulate material from filters or containers, inadequate sealing of solid sorbent sampling tubes, etc.

3. Contamination of samples due to inadequate separation of sample types or bulk sampling materials. An example is:collected airborne vapor samples shipped in the same container with bulk liquid organics.

4. Improper use of special shipping procedures designed to preserve the samples at temperatures other than ambient. This applies to those samples such as blood, ethylene oxide, and volatile pesticides, which must be shipped cold and by express carrier. Most violations will result in the need to determine the loss of integrity and a decision regarding the disposition of the sample.

9.5 Shipment of Samples

1. Check the method to determine if Packing and Shipping Instructions are included. If so, follow the instructions given.

2. Divide samples into appropriate and compatible shipping groups. Liquids will be kept separate from other materials.

3. Select appropriate shipping container and packing material.

4. Make sure that the sample is properly and accurately identified and that all necessary paperwork accompanies the shipment.

9.6 Sample Identification

A basic requirement of sample integrity is accurate sample identification. Samples that cannot be related to an associated Sample Submittal Form (Figure 9-1) because of inadequate, ambiguous, or non-existent labeling, will be discarded unless the requestor is able to provide immediate identification. Only under those circumstances where sample identification is fairly obvious, will laboratory personnel make a special effort to identify and correlate unidentified samples.

9.7 Prepare the Sample Submittal Form (Figure 9-1) as follows:

(1) Enter the sample number assigned when the sample was received and logged in. (2) The current date. (3) Enter the name of the organization and/or the individual submitting the sample for test or analysis. (4) The complete address of the person and/or the organization

submitting the sample. (5) The telephone number of the person and/or organization submitting the sample. (6) If the sample submitted is one involved in a specific project, enter the project identification. (7) Describe the sample or samples as fully as possible. (8) Indicate the source of the sample, (The organization from which it came, the location of the source, the method of collection, any ambient conditions affecting the nature of the sample, etc.) (9) The number of samples submitted with this submission. (10) The date of collection or acquisition of the sample. (11) The date the sample was shipped by the submitter. (12) The sample number. In the case of single samples this number is the same as the entry in (1) above. When more than one sample is submitted, each individual sample in the ''family'' of samples should be assigned a different number such as 1236-a, 1236-b, or 1236-1, 1236-2, etc. (13) Enter the type of sample such as ''bulk'', ''Aqu. Sol.'' ''Tinct'' ''Cores,'' ''Swatches,'' ''blood,'' ''urine,'' etc. (14) The manufacturer or producer. (15) The manufacturer's or producer's lot number. (16) Specify exactly the test wanted using ASTM or other standard number, or, using standard chemical nomenclature, the chemical analyte requested. Do not abbreviate. (17) Enter any remarks pertinent to the line entry. (18)

Enter any comments with reference to any unusual conditions detected, recommendations made, etc. (19) Note any chemicals which may be present in the workplace which may interfere with the analysis. If unsure, note the chemical anyway.

SAMPLE SUBMITTAL FORM

Sample Log No. __(1)_____ Date __(2)_____

Originator: __(3)_____

Address __(4)_____

_____

Telephone No. __(5)_____ Project No. __(6)_____

Sample Description. __(7)_____

Sampling Source. __(8)_____

Number of Samples Submitted. __(9)_____

Date of Collection __(10)_____

Date of Sample Shipment __(11)_____

Request for Analysis

| Sample Field Number | Sample Characteristics | | | Test or Analyses Requested | Remarks |
	Type	Manuf.	Lot No.		
(12)	(13)	(14)	(15)	(16)	(17)

Comments __(18)_____

_____

Possible Interfering Compounds __(19)_____

_____

FIGURE 9.1

78

10.1 Purpose

This section describes the procedures to be followed when strict chain-of-custody protocols for samples received must be followed.

10.2 Scope

This laboratory does not normally follow strict chain-of-custody procedures in handling routine samples received for testing or analysis. The submitter is usually responsible for determining whether chain-of-custody procedures are necessary for a particular sample or sample set. Usually chain-of-custody documentation is necessary when laboratory results are to be used as evidence in legal proceedings. This documentation is prepared in addition to the normal sample processing paperwork.

10.3 Chain-of-Custody Documentation Form

The Chain of Custody Documentation Form (Figure 10-1) will be used as the chain of custody record. One of these forms will be completed for each sample submitted for analysis or testing when requested by the submitter. The form is to remain with the sample at all times. The top portion of the form will be completed by the Quality Control Coordinator as follows: (1) Enter the name of the person and/or organization submitting the sample. (2) Enter the sample number from the Sample Log Sheet entry. In the case of multiple samples, give each sample within the sample set a unique sample based on the sample set number. This number must be firmly affixed in some manner to the sample or sample container. (3) Enter the submitter's address. (4) Enter the submitter's telephone number. (5) Describe the sample in sufficient detail so that it can be distinguished from other samples. (6) Indicate the sample source. (7) Record the date and time of acquisition if available. (8) Describe the method of shipment. If the sample was hand carried, so note. (9) Record the date of receipt. (10) Enter the name of the individual receiving the sample. (11) Describe the condition of the package or packages, when received noting damage, leakage or breakage and any discrepancies in labeling. (12) For bulk material samples, record the initial container weight. (13) Enter the location where the sample is stored awaiting test or analysis. Items (9) through (13) will be filled out by the Receiving Clerk.

10.4 Accountability Record

The Accountability Record portion of the form is to be completed by the laboratory technician to whom the sample is assigned for test or analysis. It establishes a record of possession of the sample while it is being processed in the laboratory. Each person handling the sample (14) records the date and time the sample is removed from its original location (15) and places his or her initials in the adjacent block. (16) The location to which the sample is removed is entered. (17) In the case of bulk samples, or those not completely consumed, record the amount returned, as a percentage of the original, (18) Record the date and time of return. (19) to whom it was given, (e.g., the name of the receiving clerk), and (20) the location where the remainder of the sample is stored. (21) Enter the date of final disposition. (22) Enter the name of the person making the final disposition. (23) Describe in detail how the final disposition was made.

CHAIN OF CUSTODY DOCUMENTATION

NAME OF SUBMITTER (1)   SAMPLE #: (2)

ADDRESS: (3)   TELEPHONE NUMBER: ___(4)___

DESCRIPTION OF SAMPLE: (5)

SAMPLE SOURCE: (6)

DATE AND TIME OF COLLECTION: (7)

METHOD OF SHIPMENT: (8)

_____

DATE RECEIVED IN LAB: _(9)_____ RECEIVED BY WHOM: ___(10)_____

RECEIPT CONDITION (CONTAINERS, PACKAGING, AND LABELING):__(11)_____

INITIAL WEIGHT OF CONTAINER (BULKS): __(12)_____

WHERE INITIALLY STORED:__(13)_____

_____

ACCOUNTABILITY RECORD

REMOVAL DATE &TIME (14)	BY WHOM (15)	LOCATION (16)	RETURNED AMOUNT (17) (% OF INITIAL)	DATE TIME (18)	TO WHOM (19)	LOCATION (20)

SAMPLE DISPOSITION:

DATE DISPOSED: (21)

BY WHOM: (22)

HOW DISPOSED:

FIGURE 10-1

11.1 Purpose

This section describes control measures necessary to identify sources of measurement error within the laboratory, and to estimate their bias and variability (repeatability and replicability)

11.2 Scope

This section prescribes the following control measures to be conducted within the laboratory:

11.2.1 Function checks will be performed by the test technician or analyst to check the validity of the sample and performance of the equipment.

11.2.2 Control checks will be performed during the analysis or testing process. These checks are used to:

a. determine the performance of the analytical or testing system.

b. quantitate the variability of results from the analysis or test in terms of precision and accuracy.

The frequency of checks will be determined by:

a. Ruggedness, precision and accuracy of the method.

b. Dependability of the technician.

c. precision and sensitivity of the instrument.

d. Difficulty and time length of the test method. Check samples will be selected from a method ''family'' at random and at irregular intervals.

11.2.3 The data from the check analyses or tests will be compared with that from the analyses or tests and tested for significant differences. Any significant differences will be reported to the Director. The Quality Control Coordinator will prepare a monthly report showing, by test method the percent of significantly different check analyses or tests that occurred during that period and will recommend corrective action.

11.2.4 Control charts will be set up and used to record results from selected function and control checks to determine when or if the testing or analytical process is out of control and to record the results of corrective action taken.

11.3 <u>Control Measures</u>

    11.3.1 Function checks are performed to verify the stability and validity of the sample and the performance of testing and analytical equipment. The testing or analytical technician will be provided with written performance specifications for each function check, accompanied by recommended action if the specifications are not met. Function checks are meausurement-method specific.

    11.3.2 Control checks will be performed during the analytical or testing process. These checks are made on all analyses or tests, and intermittently after a specified number of procedures have been completed. Some control checks may be required as part of the routine analysis or test, and are performed by the analyst or test technician to determine the performance of the system. These control checks include the use of sample blanks to observe zero concentration drift; introduction of spiked samples to determine percentage of sample recovery during intermediate extraction steps; and processing of sample aliquots to observe within- and between-run variability for the entire analysis. Other control checks are performed by the analyst or test technician intermittently to quantitate analysis variability in terms of precision and accuracy. Control samples normally used for this purpose are sample aliquots to determine precision and standard reference materials (see section 15.0) or standard reference samples to determine accuracy in addition to precision.

11.3.3 Results from selected function checks and control samples will be recorded on control charts: to track events showing that the system is out of control; to indicate what part of the system is the source of the error; and to provide an indication of the results of any corrective action taken. The Quality Control Coordinator is responsible for the establishment and monitoring of any control charts needed for laboratory performance control.

	Effective Date	Subject	Section No. 12.0
	By	QUALITY DOCUMENTATION AND RECORDS CONTROL	Page 1 of 5
	Approved		Revised

12.1 <u>Purpose</u>

This section describes how the laboratory will control the issue and retrieval of all documents relating to the analytical and testing activities of the laboratory, to the control of quality of these activities and to the storage and security of the technical documentation generated by these activities.

12.2 <u>Scope</u>

The controls described in this section are limited to the documents and reports listed below, and do not affect documentation related to other laboratory management activities. Such document control procedures are described in other laboratory standard operating procedures.

12.3 The most important elements of the quality assurance program to which document control is applied include:

- Sampling procedures
- Calibration procedures
- Analytical and test procedures
- Data collection and reporting procedures
- Auditing procedures
- Computational and data validation procedures
- The Quality Assurance Manual
- Analytical and testing reports
- Laboratory notebooks

- Validation reports

- Vendor and internal audit reports

- Calibration and preventive maintenance records

- Chain-of-custody documentation

- Technical document change requests

- Corrective Action Requests

12.4 The quality records listed above which are originated and maintained as hard copies will be retained for a period of five years in laboratory files. After a period of five years, they will be reviewed for disposition. Those not discarded will be transferred to microfiche and retained indefinitely. Records generated by computer will be retained in that format indefinitely, with due care to keeping them in safe, hazard-free storage.

12.5 The Quality Control Coordinator will maintain full control over the distribution of documents listed in Par. 12.3, above. A file control will be maintained showing the following information:

Document number	Distribution
Title	Format
Originator of the document	File location
Latest issue date	Length of retention
	Change number

Signature of the persons receiving the document.

Whenever a change is made to a controlled document, the Quality Control Coordinator will issue the new document together with the Technical Document Change Notice (Figure 12-1). The Technical Document Change Notice is to be filled out as follows: (1) The date of preparation of the Change Notice. (2) The Technical Document Change Notice number. This number should also be entered on the document being changed. (3) The effectivity point should be clearly established, advising the recipient of the cut-off point when the old document becomes obsolete and the change becomes effective. The effectivity point can be described in terms of a date, the receipt of a new contract, the depletion of a supply of test samples, etc. (4) Indicate the type of document being changed. (5) Enter the document title. (6) Identify the individual initiating the Change Notice. (7) Describe the condition or information that is to be changed. (8) Describe, in detail, the nature of the change to be made. (9) Give in detail, the reasons for making the change. (10) List the individuals or laboratory sections or departments to whom the changed document is to be distributed. (11) When the changed document is distributed, obtain the signature of each recipient. (12) List the costs, if any, involved in accomplishing the change described by the Notice. Give details on the reverse side of the Change Notice. (13) Indicate here all necessary approvals obtained in order to ratify the Change Notice.

The Quality Control Coordinator is responsible for seeing that copies of obsolete documents are removed from points of use and are destroyed. He is also responsible for monitoring activity to ensure that approved changes are incorporated into the laboratory's routine work activity.

12.5 Document Changes

12.5.1 A Request for changes to methods, sampling data sheets, or calibration instructions may be made by anyone, the request being made in writing on the Technical Document Change Notice, since this form acts as a request when being initiated. It must be approved by the Laboratory Director before the change is published and distributed.

12.5.2 Changes may be promulgated by the issue of entire new documents, of replacement pages thereto, or, in the case of corrections of errata, etc., by pen and ink posting on the original document with this action noted on the Change Notice.

12.6 The Quality Control Coordinator is responsible for distribution and retrieval of documents and for obtaining all required signatures.

12.7 All files, when not in use will be kept in locked file cabinets. Computer data security will be maintained in accordance with data processing standard operating procedures for restricting entrance to computer data information.

XYZ LABORATORIES, INC.

TECHNICAL DOCUMENT CHANGE NOTICE

DATE _____(1)_____

TDCN No. _____(2)_____

EFFECTIVE _____(3)_____

(4)
DOCUMENT
TYPE
_____ METHOD

_____ SAMPLING DATA SHEET

_____ CALIBRATION INSTRUCTION

_____ OTHER (DESCRIBE)

DOCUMENT TITLE _____(5)_____

REQUESTED BY: _____(6)_____

CHANGE FROM: _____(7)_____

CHANGE TO: _____(8)_____

REASON FOR CHANGE: _(9)_

DISTRIBUTIONS	SIGNATURES	*COSTS (12)	s	APPROVALS: (13)
(10)	(11)	NEW EQUIP.		
		NEW MAT'L		
		SCRAPPED OR OBS. EQUIP.		
		SCRAPPED OR OBS. MAT'L		
		PERSONNEL		
		OTHER		

* DETAIL ON REVERSE SIDE

FIGURE 12-1

13.1   Purpose

This section is written to assure that only gages and instruments which are properly and currently calibrated are used in making determinations, the results of which are recorded and reported upon.

13.2   Scope

The calibration of laboratory instruments falls into two categories: calibration which is conducted on a routine basis prior to each use; and periodic, scheduled calibration of instruments and gages against known standards, traceable to the National Institute of Science and Technology (NIST) to ensure the continuing precision and accuracy of such instruments. The calibration policies and procedures set forth in this section apply to all instruments in the latter category, including:

a. Sampling equipment at remote sampling stations

b. Analytical and test equipment in the laboratory

c. Flow rate (e.g., rotameters), volume (e.g., dry gas meters), pressure, vacuum and temperature measurement equipment, balances, hardness testers, tensile testers, ph meters, electrical meters, etc.

13.3   Calibration Standards Quality

Transfer standards will have 4 to 10 times the accuracy of field and laboratory instruments and gages. For example, a thermometer used in the field to determine air temperature, having a specified accuracy of $\pm 2°F$, will be calibrated against a laboratory thermometer with an accuracy of $\pm 0.5°F$. The Standards used in the laboratory measurement system will be calibrated against higher-level, primary standards having unquestionable and higher accuracy. These higher-level standards will be certified by NIST or other recognized standardization bodies.

Calibration gases purchased from commercial vendors will be required to have a certificate of analysis. Whenever a certified, calibration gas is available from NIST, commercial gas vendors will be required to establish traceability of the certificate of analysis to the certified gas. Since inaccurate concentrations of certified gases may result in serious measurement errors, the Quality Control Coordinator will spot-check the accuracy and validity of certification of calibration gases as a part of the Receiving Inspection process. (See Section 8.0.)

13.4 <u>Environmental Conditions</u>

Measuring and test equipment and calibration standards will be calibrated in an area that provides control of environmental conditions to the extent necessary to assure required accuracy and precision. The calibration area will be kept reasonably free of dust, vapor, vibration and radio frequency interferences; and it will be remote from or shielded from equipment producing noise, vibration, chemical or micro-wave emissions, etc. The laboratory calibration area will have temperature and humidity controls, maintaining a temperature of 68°F to 73°F and a relative humidity of 35% to 50%. A filtered air supply will be provided, and hoods will be checked for proper operation before each use.

13.4.1 <u>Electric Power</u>. The power for the laboratory will be regulated to provide a continuous voltage of ±5%; control of harmonic distortion; control of line transients caused by interaction of other users; and a suitable grounding system to assure equal potentials throughout the laboratory.

13.4.2 <u>Lighting</u>. Adequate lighting at levels of 80 to 100 foot-candles at work bench areas. Fluorescent lights will be shielded to reduce electrical noise.

13.5 <u>Calibration Intervals</u>.

All calibration standards and instruments and gages will be assigned an established calibration interval. The form for listing gage and instrument intervals is Figure 13-1 which is filled out as follows: (1) List all instruments and gages which are included in the laboratory calibration system, by instrument type or family. Instruments should not be listed individually. (2) Enter the calibration interval for each instrument in terms of elapsed time or number of completed cycles of use. (3) Enter the Calibration Procedure or method number which pertains to the instrument group.

In the absence of an established interval based on the equipment manufacturer's recommendations, an initial calibration period will be assigned by the Quality Control Coordinator. The calibration intervals will be adjusted from time-to-time, based on experience gained through use over a period of time, as evidenced by data from gage calibration records. The calibration intervals will be specified in terms of time or, in the case of certain types of measuring and test equipment, usage cycles, or usage periods.

The choice of significant intervals will be based on the inherent stability and sensitivity of the instrument, its purpose or use, accuracy, conditions of use, and frequency or amount of usage. The intervals will be shortened or lengthened after evaluating the results of present and past calibrations and adjusting the schedule to reflect the findings. These evaluations must provide positive assurance that calibration interval adjustments will not adversely effect the accuracy of the system.

Adherence to the Calibration Frequency Schedule is mandatory. Compliance is assured by the following procedure: For each gage or instrument in the calibration system, a Gage Calibration Record card is prepared as follows: (1) Enter the gage or instrument assigned calibration number. In this laboratory, the calibration number is the same as the Property Inventory Number. (2) Enter the description of the gage or instrument. (3) Give the manufacturer's name. (4) Enter the equipment serial number. (5) Enter the Model Number. (6) Indicate the location of the equipment or the department or section where used. (7) Enter the calibration frequency from the Calibration Frequency Schedule (Figure 13-1). (8) Enter the applicable Calibration Procedure number. For each scheduled calibration operation enter: (9) the date that the calibration is performed; (10) the name of the individual performing the calibration; (11) the section or department using the instrument or gage at the time of calibration; (12) The results of the calibration.

The Gage Calibration Record Cards will be sorted and filed by month, so that the cards which are in a particular month group will identify which instruments are to be recalled for calibration during that month. Notice of recall will be given prior to the due date. It may be necessary to calibrate between normal calibration dates if there is evidence of inaccuracy or damage. Calibration of a measurement device can be requested at any time, regardless of the calibration due date for that item, following the occurrence of any event that places the device's accuracy in doubt.

The Quality Control Coordinator will maintain surveillance by periodically and randomly auditing for compliance with the calibration control and recall system.

13.6 Calibration Procedures

Written step-by-step procedures for calibration of measuring and test equipment, and for the use of calibration standards will be used in order to eliminate possible measurement inaccuracies due to differences in techniques, environmental conditions, choice of higher-level standards, personnel changes, etc. These calibration procedures may be prepared by the laboratory, may be published standard practices or written instructions that accompany purchased equipment, or may be obtained from the Government-Industry Data Exchange Program Metrology Data Bank (GIDEP). These procedures will contain the following information:

13.6.1 The specific equipment or group of equipment to which the procedure is applicable. Like equipment or equipments of the same type, having compatible calibration points, environmental conditions, and accuracy requirements, may be serviced by the same calibration procedure.

13.6.2 A brief description, or abstract, of the scope, principle, and/or theory of the calibration method.

13.6.3 Fundamental calibration specifications, such as calibration points, environmental conditions, and accuracy requirements.

13.6.4 A list of calibration standards and accessory equipment required to perform the calibration steps. The manufacturer's name, model number, and required instrument accuracy will be furnished as applicable.

13.6.5 The complete procedure for performing the calibration, arranged in a step-by-step manner, clearly and concisely written.

13.6.6 The calibration procedures will provide specific instructions for obtaining and recording the calibration data, and will include, where applicable, copies of calibration data recording forms.

Note: Many calibration procedures require statistical analysis of results.

13.7 Calibration Sources

All calibrations performed by the laboratory will be traced through an unbroken chain, supported by reports or data sheets to ultimate or national reference standards maintained by a national organization such as The National Institute of Science and Technology (NIST). The laboratory may also use, at its discretion, as an ultimate reference standard, an independent, reproducible standard, such as a standard that depends on accepted values of a natural physical constant. An up-to-date calibration report for each standard used in the laboratory will be retained in the laboratory's files. When calibration services are provided by an outside metrology laboratory, on a contract basis, copies of the calibration reports furnished by them will also be kept on file. All standards calibration reports will contain the following information:

a. Report number

b. Identification number of the calibration standard to which the report pertains.

c. Environmental conditions under which the calibration was performed.

d. Required accuracy of the calibration standard.

e. Deviation or corrections.

f. Corrections that must be applied if standard conditions of temperature, etc, are not met or differ from those at the place and time of calibration.

Contracts let for calibration services will require the metrology laboratory to furnish records on the traceability of their calibration standards.

13.8 Labeling

All equipment to be calibrated will have affixed to it in plain sight, on the instrument itself, or its container, a tag or label (Figure 13-3). This label is to be filled out in the following manner: (1) Enter the name of the person performing the calibration service. (2) Enter the date the calibration was performed. (3) Enter the date that the next calibration service is due, based on information from the Gage Calibration Record Card (Figure 13-2) and the Calibration Frequency Schedule (Figure 13-1). Equipment past due for calibration will be sequestered, or if this is impractical, will be impounded by tagging with the Out of Service label (Figure 13-4) which will be filled out as follows: (1) Enter the signature of the Quality Control Coordinator. (2) Enter the date impounded or otherwise taken out of service. Certain instruments in the laboratory are used only as indicators, and the readings from them are not recorded or reported. These instruments will be labeled with the Not Calibrated label (Figure 13-5).

CALIBRATION FREQUENCY SCHEDULE

Type	Minimum Frequency	Instruction Number or Calibration Method
(1)	(2)	(3)

Figure 13-1

INSTRUMENT / GAUGE CALIBRATION RECORD

INSTRUMENT No. (1)	DESCRIPTION (2)	IDENTIFICATION No. (1)
MFR. (3)	SERIAL No. (4)	MODEL No. (5)
LOCATION (6)	CALIBRATION FREQUENCY: EVERY _____ (7)	CALIBRATION PROCEDURE No. (8)

DATE CALIBRATED	CHECKED BY	DEPT.	RESULTS
(9)	(10)	(11)	(12)

(OVER)

DATE CALIBRATED	CHECKED BY	DEPT.	RESULTS

FIGURE 13-2

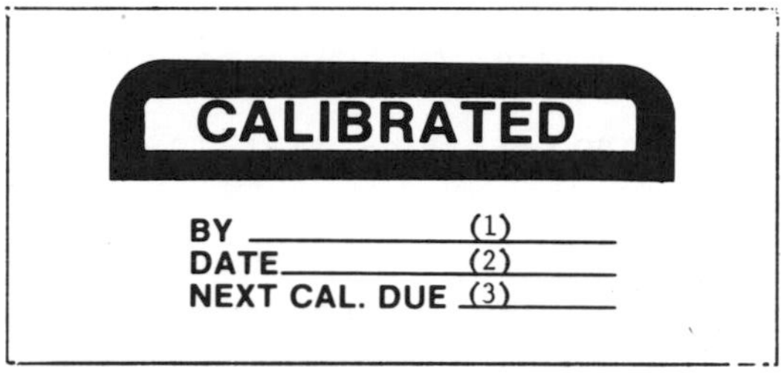

CALIBRATED

BY _____ (1) _____
DATE_____ (2) _____
NEXT CAL. DUE (3) _____

FIGURE 13-3

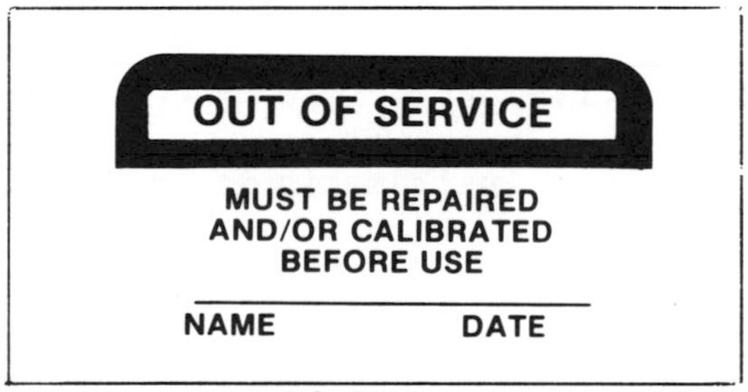

OUT OF SERVICE

MUST BE REPAIRED
AND/OR CALIBRATED
BEFORE USE

NAME   DATE

FIGURE 13-4

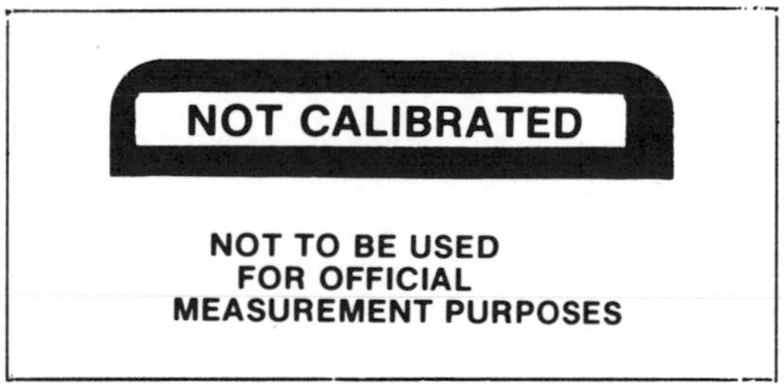

NOT CALIBRATED

NOT TO BE USED
FOR OFFICIAL
MEASUREMENT PURPOSES

FIGURE 13-5

	Effective Date	Subject	Section No. 14.0
	By	PREVENTIVE MAINTENANCE	Page 1 of 3
	Approved		Revised

14.1 Purpose

This section describes the Quality Control Coordinator's responsibilities for carrying out a comprehensive preventive maintenance program for gages, instruments, and test equipment used in its analytical and testing activities.

14.2 Scope

This section applies to all gages, measuring equipment, and test instruments included in the laboratory calibration program; i.e., any equipment used for making determinations, the results of which are recorded and reported upon.

14.3 Preventive Maintenance

The laboratory will conduct an orderly program of positive actions (equipment cleaning, lubricating, reconditioning, adjusting, and/or testing) to prevent instruments or equipment from failure during use. The purpose of this preventive maintenance program is to increase measurement system reliability and thus increase data completeness. Conversely, if the preventive maintenance program is not properly carried out the results will be: increased measurement system downtime with a decrease in data completeness; increased maintenance costs and a distrust in the validity of the data. Data completeness is a criterion used to validate data.

14.3.1 <u>Preventive Maintenance Schedule</u> The Quality Control Coordinator will prepare and implement a preventive maintenance schedule for laboratory measurement systems. The planning required to prepare the schedule will take into consideration:

14.3.1.1 Highlighting that equipment or parts thereof which are most likely to fail without proper preventive maintenance.

14.3.1.2 Defining a spare parts inventory which should be maintained to replace worn-out parts with a minimum of downtime.

The preventive maintenance schedule should relate the purpose of analysis or testing, environmental influences, physical location of equipment, and the level of operator skills. Checklists will be used to specify maintenance tasks, as recommended by manufacturers of the equipment. Frequencies or time intervals between maintenance service will also be specified in accordance with recommended practices.

Since instrument calibration is commonly the responsibility of the operator in addition to preventive maintenance tasks, a combined preventive maintenance-calibration schedule will be used in those cases.

14.3.2 <u>Preventive Maintenance Record</u> A record of all preventive maintenance and daily service checks will be maintained by the Quality Control Coordinator. Daily service checklists will be filed with the measurement data.

Effective Date	Subject	Section No. 15.0
By	REFERENCE STANDARDS	Page 1 of 3
Approved		Revised

15.1 Purpose

This section discusses the use of Standard Reference materials
available from the National Institute of Science and Technology
(NIST).

15.2 Scope

This section deals with the Standards Reference Materials used in
analyses and tests. It does not concern itself with the measurement
standards discussed in Section 13.0, ''Control of Measuring and
Test Equipment.''

15.3 Availability of Standard Reference Materials

The Quality Control Coordinator will keep on file a current copy of
NIST Special Publication 260, ''NIST Standard Reference Material
Catalog'' for the information of all concerned. As an example, a
copy of the NIST (formerly NBS) Certificate of Analysis No. 2676.
(Figure 15-1) is shown here. This consists of three concentrations
of lead, cadmium, zinc, and manganese on membrane filters. This
illustrates a typical SRM available from NIST.

15.4 Use of SRM's

This laboratory will use Standard Reference Materials as available
from NIST for calibration of instruments, preparation of lower
tier standards or as reference standards against samples to be
analyzed or tested as applicable.

U. S. Department of Commerce
Rogers C.B. Morton
Secretary

National Bureau of Standards
Richard W. Roberts, Director

National Bureau of Standards
Certificate of Analysis
Standard Reference Material 2676
Metals on Filter Media
(Pb, Zn, Cd, Mn)

This Standard Reference Material is intended primarily for use as an analytical standard for the determination of cadmium, lead, manganese, and zinc in the industrial atmosphere. The SRM consists of a set of membrane filters on which have been deposited the indicated quantities of salts of the particular metals.

Filter	Metal Content, µg/filter			
	Cd	Pb	Mn	Zn
A 1	0.50 ± .04	6.8 ± 1.1	1.93 ± .29	1.02 ± .06
A 2	2.48 ± .14	29. ± 2.6	10.3 ± 1.5	5.10 ± .26
A 3	10.1 ± .4	102. ± 6	20.6 ± 1.0	10.1 ± 1.1

The filters were prepared by depositing on them known and carefully reproduced volumes of solutions of pure salts, using the technique described in NBS report NBSIR 73-256.

The certified values are based upon determination of the metal content by atomic absorption spectrometry and by polarographic measurement. In these analyses, entire filters were mineralized by digestion in acid prior to measurement. The certified values are the means of those found by the two techniques while the uncertainties represent the 95 percent tolerance limits based on measurement error and variability between samples*.

The filters are identified by the numbers A1, A2, A3 printed on their edge. The metal content of the inked identification is negligible so it need not be removed. An entire filter must be used for each measurement since the metals are not uniformly distributed.

* See page 14, The Role of Standard Reference Materials in Measurement System, NBS Monograph 148, 1975. The concept of tolerance limit is also discussed in Chapter 2, Experimental Statistics, NBS Handbook 91, 1966.

In brief, if measurements were made on all the units, almost all (at least 95 percent) of these measured values would be expected to fall within the indicated tolerance limits with a confidence coefficient of 95 percent (or probability = .95).

(over)

FIGURE 15-1

The filters were prepared by R. Mavrodineanu and J. R. Baldwin. Atomic absorbtion analyses were made by them and also by T. C. Rains. E. J. Maienthal performed the polarographic analyses.

The overall direction and coordination of the technical measurements leading to certification were under the chairmanship of J. K. Taylor.

The technical and support aspects involved in certification and issuance of this Standard Reference Material was coordinated through the Office of Standard Reference Materials by W. P. Reed.

Washington, D.C. 20234   J. Paul Cali, Chief
June 30, 1975   Office of Standard Reference Materials

16.1 <u>Purpose</u>

This section explains the need for data validation and the methods of data validation which will be employed by this laboratory.

16.2 <u>Scope</u>

Data validation can be accomplished by several methods. The validation process can be manual or computerized.

16.2.1 Data validation is the process in which data are checked and accepted or rejected based on a set of criteria. Validation is performed to isolate spurious values since such values are not automatically rejected. Records of invalid data found will be retained in accordance with established records retention policies.

16.2.2 Validation methods will include review by supervisors as well as comparison with criteria by computer. Criteria will depend on the types of data and on the purpose of the measurement.

16.2.3 Various statistical techniques are useful. Periodic checking of manually reduced data is imperative. The statistical technique that is most commonly used is acceptance sampling.

16.3 Data Validation—General

Data validation is the process in which data are checked and accepted or rejected based on a set of established criteria. This involves a critical review of a body of data in order to identify spurious values or outlying observations. The review may be only a quick scan to detect extreme values or a detailed evaluation requiring the use of a computer. In either situation, when a spurious value is located, it is not immediately rejected. Each questionable value or anomaly must be checked for validity. Records of values that are judged to be invalid or are otherwise doubtful will be retained. These records become a useful resource of information for judging data quality.

16.4 Manual Data Validation

Both the technician and the laboratory supervisors will inspect integrated daily and weekly results for questionable values. This kind of inspection is most useful for the detection of extreme values or outliers appearing as unusually high or low data points.

   The criteria for determining extreme values are derived from prior data obtained from similar tests or analytical work. The data used to determine extremes may be the minimum and maximum data points for all prior data from similar tests or analyses. The time spent on checking data that have been manually reduced by technicians depends on the time available and in the demonstrated abilities of the technicians.

16.5 Computerized Techniques

As necessary, the computer will be used not only to store and retrieve data but also to validate data. The use of the computer allows the fine-tuning of the criteria for extreme values so as to be specific for individual hours during the day, if desirable. For instance, with this refinement, an hourly average concentration of carbon monoxide of 15 ppm might not be considered an extreme value at 8:00 AM, but would be evaluated as questionable if it appeared at 2:00 AM.

Another indication of spurious data is a large difference in concentrations reported in two successive time periods. The difference in concentrations, which might be considered excessive, may vary from one contaminant to another and quite possibly may vary from one sample source to another for the same contaminant. This difference will be tested for significance through statistical analysis, using an appropriate test for significance of differences, depending on the kind of information available in the data package in question. The criteria for what constitutes an excessive change may also be linked to the time of day and contaminant relationships. For instance, high concentrations of SO_2 and O_3 cannot co-exist, and such data should be suspected.

16.6   Tools Used in Data Validation

16.6.1   Strip Chart Data

When a continuous monitoring device is used, the output is an analog trace on a strip chart. Strip charts will be cut at weekly intervals and will be sent to the laboratory supervisor for interpretation. Reading strip charts is a tedious task, subject to varying degrees of error, therefore a procedure for maintaining desired quality for data manually reduced from strip charts will be followed. The procedure to be used for checking the validity of the data reduced by a technician is to have another technician or supervisor check the data. Because the values will have been taken from the strip chart by visual inspection, some differences in the values derived by two different indidivuals can be expected. When the difference exceeds a specified amount, and the initial reading has been determined to be incorrect, an error will be noted. If the number of errors exceeds a predetermined number, all data from the strip chart are rejected, and the chart is read again by another individual. The question of determining how many values will be checked can be answered by using acceptance sampling techniques.

16.6.2 <u>Application of Acceptance Sampling Techniques</u>

Acceptance sampling will be used in data validation to determine the number of data items (individual values on a strip chart) that must be checked to determine with a given confidence level that all data items are acceptable. The supervisor needs to know, with a high degree of assurance, without checking every data value, whether a defined error level has been exceeded. From each strip chart with N data values (the population) the supervisor can randomly inspect a sample of n data values. If the number of erroneous values is equal to or less than c, the rejection criteria, the values for the strip chart are accepted. If the number of errors is greater than c, the values on the strip chart are rejected and another individual is asked to read the chart. The explanation for determination of

sample sizes and for the use of rejection and acceptance

criteria appears in ANSI/ASQC Z 1.4-1981, ''Sampling

Procedures and Tables for Inspection By Attributes.''

17.1 Purpose

This section will describe the environmental controls required for laboratory operations.

17.2 Scope

This section will deal with the control of environmental conditions in laboratory working areas. Environmental controls for the calibration area are covered in Section 13.0, Par. 13.4.

17.3 Environmental Controls in Working Areas

The environment in the working areas of the laboratory will be controlled only to the extent afforded by normal, commonly used heating, ventilating, and air-conditioning equipment which will maintain a working temperature of 68° to 76°F and a relative humidity at about 50%. This laboratory has no requirement for clean-room facilities, nor are there any special requirements beyond normal good housekeeping practices.

17.4 Hoods

Hoods will be checked for proper operation prior to each use.

18.1   <u>Purpose</u>

This section sets forth the procedures and responsibilities for handling customer complaints and negative audit results.

18.2   <u>Scope</u>

This section applies to all technical complaints regardless of the source.

18.3   <u>Procedure</u>

All technical complaints and negative comments or suggestions from customers, government agencies or other sources outside the laboratory will be turned over to the Quality Control Coordinator for review, handling, and reply. In each case, he will advise the individuals concerned as to the nature of the complaint. Additionally, he will initiate corrective action measures when necessary. Upon completion of corrective action and the finding of a solution to the problem, the Quality Control Coordinator will advise the customer accordingly. In the case of corrective action taken to satisfy the comments or suggestions of outside auditors from accrediting organizations, detailed explanations will be given of measures taken to prevent recurrence of problems causing the negative comments.

19.1 Purpose

This section describes the requirements for the control of the quality of work imposed upon outside laboratories doing analytical or testing tasks which are beyond the capabilities of this laboratory.

19.2 Scope

This section applies not only to outside laboratories doing analytical or testing work on a contract basis, but also to metrology laboratories providing calibration services in accordance with Section 13.0.

19.3 Quality Assurance in Contract Laboratories

Each contract laboratory, which this laboratory employs for providing testing services, chemical analyses, or calibration services, will maintain its own internal quality assurance system. The capability of the contractor to maintain a high level of quality of work will be taken into consideration as a part of the contract evaluation process and will be weighed heavily in that process.

The Quality Control Coordinator will furnish each new contract laboratory, prior to the completion of the contract agreement, ''audit'' samples to help in gaining an understanding of the quality of data produced by the contract laboratory. The data from these samples will serve a number of purposes. First, the data may be used to help making decisions as to the ability of an outside laboratory to perform to desired standards. Secondly, the data may be used to point out problem areas in methods such as ambiguity of language, sample instability, or critical analytical or testing steps.

In furtherance of this, when a new method comes into question, the Quality Control Coordinator will furnish the contractor with a set of ''user check'' spiked samples to analyze by the method in question. This serves the dual purpose of testing the contractor's proficiency in performing the method and also testing the clarity and technical proficiency of the method language.

20.1 <u>Purpose</u>

To set forth the training methods, evaluation and qualification procedures, and responsibilities for motivational programs in the laboratory.

20.2 <u>Scope</u>

All personnel involved in any function effecting data quality (sample collection, analysis, testing, data reduction, and quality control and assurance), will have sufficient training in their appointed positions to contribute to the reporting of complete, high quality data. The Quality Control Coordinator is responsible for seeing that the required training is made available to these personnel and that records are maintained on each person reflecting satisfactory completion of training programs or qualification tests.

20.3 <u>Quality Training Objectives</u>

Quality Control training programs will have objectives seeking solutions to laboratory quality problems. These training objectives will help to develop for all laboratory personnel involved in any aspect or function affecting quality, those attitudes, that knowledge, and those skills which will enable each one to contribute to the production of high quality data continuously and effectively.

20.4 <u>Available Training Methods</u>

    20.4.1 Experience training—This is on-the-job training (OJT), learning to cope with problems using prior experience or knowledge as the basis for action.

    20.4.2 Guidance training—This is OJT with outside help from supervisors or co-workers. The advice may be solicited, provided on an informal basis, or on a planned, scheduled basis. On-the-job training will be conducted as follows:

        a. Observe an experienced operator perform the different tasks in the measurement process.

        b. Perform the operation under the direct supervision of an experienced operator.

        c. Perform the operations independently but with a high level of quality control checks using the techniques described in Par. 20.5 below, which deals with operator proficiency evaluation procedures.

    20.4.3 Independent study—This may be night school classes, outside reading, attendance at seminars, or professional society meetings, etc. on a voluntary basis. (See Personnel Department Policy No.--- for information on tuition and fee payments by the laboratory.)

20.4.4 In-house training—This is class room study, held during working hours, presented on a formal basis. Courses may be given by qualified laboratory personnel or outside instructors.

20.4.5 Outside part-time courses—It is the policy of the laboratory to encourage continued education in all job related areas. There are many sources of quality control training available. The Quality Control Coordinator will seek out and maintain lists of schools, professional societies, and other organizations offering courses related to laboratory quality control activities. He will be available for consultation with those seeking continuing education opportunities in the field of quality control.

20.5 Training Evaluation

Training will be evaluated in terms of: (1) level of knowledge and skill achieved by the operator from the training, and (2) the overall effectiveness of the training including determination of the training areas which need improvement. When a quantitative performance rating is made on the operator during the training period, in terms of skill and knowledge achieved, this rating will also provide an assessment of the overall effectiveness of the training program.

Several techniques are available to evaluate the operator and the effectiveness of the training program. One or more of these techniques will be used during the evaluation. The most common types of evaluation techniques applicable to a measurement system training program are the following:

20.5.1 Testing—A written test following training will be given in order to assess the effectiveness of the training effort. The instructor may, because of circumstances, give a pre-training test in order to determine the level of knowledge prevailing prior to the start of the training course. Post-training tests also provide the instructor with information on training areas which need improvement.

20.5.2 Proficiency checks—In order to measure skill improvement in both OJT and short course training, trainees will be assigned work tasks related to training subject matter. Accuracy and completeness will be used as indicators to score the trainees' proficiency. The work tasks will be in the following form:

a. Sample collection—Trainee will be asked to list all steps involved in sample collection for a hypothetical case. In addition, the trainee will be asked to perform selected calculations.

b. Testing or analysis—Trainee will be provided unknown samples to which a prescribed method is to be applied. As used here, an unknown is a sample whose composition or identity is known to the supervisor (OJT) or instructor (of a course) but unknown to the trainee. Proficiency is judged in terms of accuracy.

c. Data reduction—Trainees responsible for data reduction will be given data sets to validate. Proficiency will be judged in terms of completeness and accuracy.

When proficiency checks are conducted on a recurring basis, a control chart will be used to show progress of the training effort. The results of all completed training courses will be posted to the individuals' personnel jacket. Note will be made of the expiration dates of qualification certificates, so that a follow-up can be made to ensure timely renewal.

20.5.3 Interviews

As necessary, oral interviews will be used to determine whether training efforts were effective. Interviews may be conducted by the immediate supervisor or the Quality Control Coordinator, as appropriate to the situation.

20.6 <u>Quality Motivation</u>

The Quality Control Coordinator is responsible for conducting,

from time to time, such motivational campaigns as deemed necessary.

He will also make recommendations to management with regard to

efforts to increase employee awareness of each individual's

responsibility for the quality of laboratory output.

21.1 <u>Purpose</u>

This section lists a number of statistical tools and techniques
available to the laboratory worker. These tools and techniques are
used to gain more information about data produced by testing and
analytical procedures. This information will be used to make
predictions or come to conclusions about the nature of the data.

21.2 <u>Scope</u>

The list given here is provided to familiarize the user with the
kind of statistical tools that may be required for use by regulatory
or accrediting bodies, or may be otherwise useful to the
laboratory. This section will not provide detailed instructions
for the use of such tools or techniques.

21.3 <u>Statistical Tools and Techniques</u>

   a. Summary statistics—Summary statistics such as the mean ($\overline{X}$), the
      standard deviation (σ) and the variance (σ^2) are used to
      simplify the presentation of data and at the same time summarize
      essential characteristics.

   b. Frequency distributions—Frequency distributions such as normal
      and log-normal distributions are used to present relatively
      large data sets. Their understanding and use is important
      because the usefulness of control charts is based on the
      assumption that the raw data is normally distributed. For data
      that is log-normally distributed, such as the daily

concentrations of suspended particulates in ambient air, transformation provides the normal distribution used in making decisions concerning the results.

c. Tests of hypotheses—A hypothesis is a belief held or an assertion made about the nature of numerical values of some parameter of a population. In making an hypothesis there is the opportunity of making one of two kinds of errors:

(1) Type I error which is rejection of the hypothesis when it is true.

(2) Type II error which is acceptance of the hypothesis when it is false.

d. Outliers—Outlying observations are unusually large or small values which appear as anomalies in a data set. A number of statistical tests are available for determining the presence of outlying data.

e. Control Charts—Control charts are, perhaps, the most useful statistical tool available to the laboratory. They will be used in this laboratory as prescribed by methods and as directed by the Quality Control Coordinator (Figures 21-1 and 21-2).

f. Sampling—This laboratory will use sampling plans from ASQC Z 1.4-1981, ''Sampling Procedures and Tables for Inspection By Attributes'' for incoming inspection of testing and analytical supplies and materials as directed by the Quality Control Coordinator, and for the selection of sample data for data validation.

g. Tests for significance of difference—Differences often appearing between two sets of data give rise to the question of whether the difference has any importance or significance, or occurs merely due to chance cause. In these cases, tests using the F, t, or Chi-Square distributions are useful in making a decision about the data sets in question.

h. Reliability and maintainability—As measurement systems grow more complex, system reliability becomes an increasingly important concern in determining the completeness and accuracy of results. This is especially true when dealing with remote recording instruments.

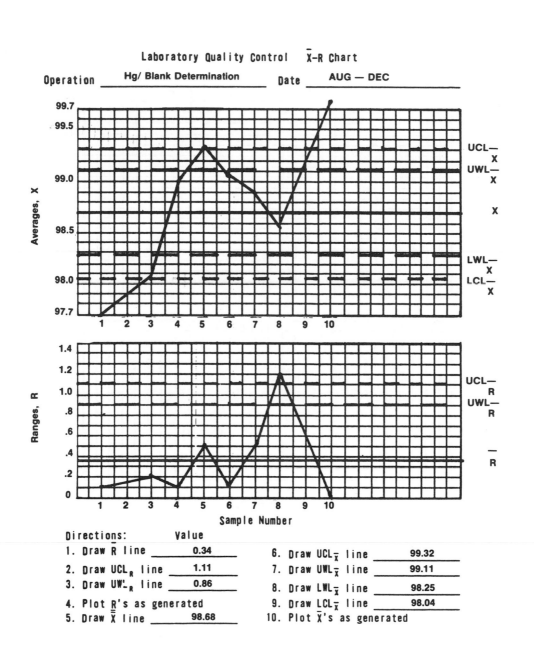

Laboratory Quality Control X̄-R Chart

Operation ___Hg/ Blank Determination___ Date ___AUG — DEC___

Directions: Value
1. Draw R̄ line ___0.34___   6. Draw UCL_X̄ line ___99.32___
2. Draw UCL_R line ___1.11___   7. Draw UWL_X̄ line ___99.11___
3. Draw UW'_R line ___0.86___   8. Draw LWL_X̄ line ___98.25___
4. Plot R's as generated   9. Draw LCL_X̄ line ___98.04___
5. Draw X̄̄ line ___98.68___   10. Plot X̄'s as generated

FIGURE 21-1

129

Laboratory Quality Control $\bar{X}$–R Chart

Operation _____   Date _____

Averages

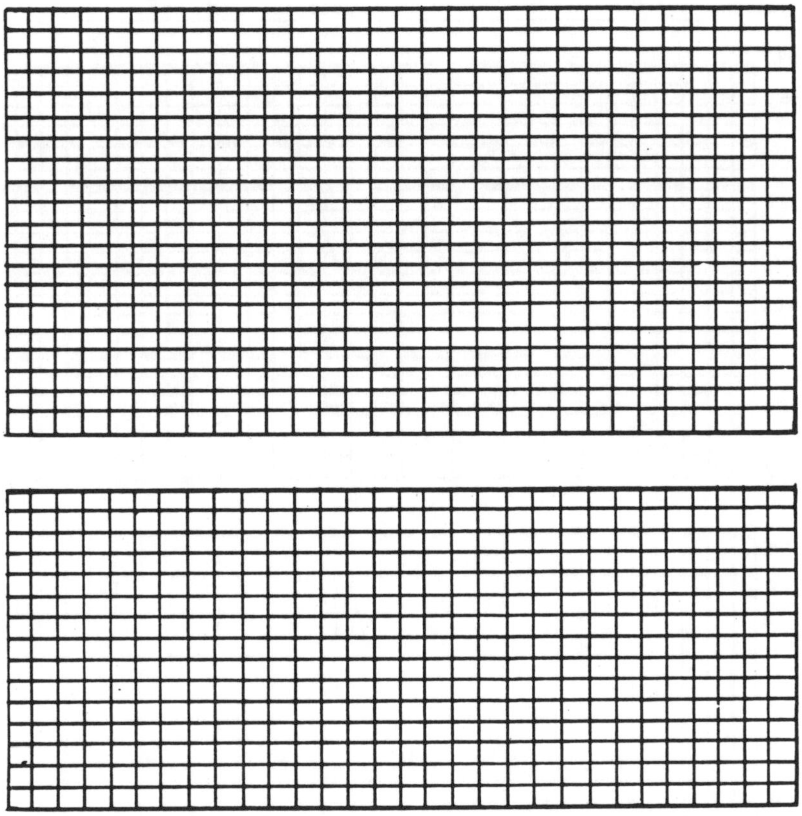

Ranges

Sample Number

Directions: Value
1. Draw $\bar{R}$ line _____
2. Draw UCL_R line _____
3. Draw UWL_R line _____
4. Plot R's as generated
5. Draw $\bar{\bar{X}}$ line _____

6. Draw $UCL_{\bar{X}}$ line _____
7. Draw $UWL_{\bar{X}}$ line _____
8. Draw $LWL_{\bar{X}}$ line _____
9. Draw $LCL_{\bar{X}}$ line _____
10. Plot $\bar{X}$'s as generated

FIGURE 21-2

130

22.1 <u>Purpose</u>

The purpose of this section is to prescribe actions to be taken when there is suspicion of nonconforming results occurring during testing or analytical operations, or when such nonconformities occur as a result of data validation, in independent audit reports, in customer complaints, or as a result of equipment malfunction.

22.2 <u>Scope</u>

The subject material in this section is related to that of the following Section 23.0—''Corrective Action.'' However, Corrective Action is the most important one of the several steps that should be taken when a nonconformity is discovered, therefore, it is treated separately.

22.3 <u>Actions</u>

22.3.1 When a nonconforming result is discovered or suspected, the occurrence will be reported immediately and the details of the event noted in the laboratory notebook.

22.3.2 The Quality Control Coordinator will be responsible for reviewing the circumstances of all instances of occurrence of nonconformities, to determine whether corrective action should be taken, if new samples are required, if the customer should be notified, if retesting or reanalysis is necessary, or whether the results should be confirmed by independent third party testing or analysis. The results of this review by the Quality Control Coordinator will be recorded on the Nonconformance Event Review Report (Figure 22-1) in which the first 6 entries will be filled out by the analyst or test technician as follows: (1) Enter the Sample Identification Number. (2) Enter the date the Report is originated. (3) Describe the sample in as much detail as possible. (4) Enter the number and name of the method being used at the time of the occurrence of the nonconformity. (5) Describe the nonconforming event suspected. (6) Describe the expected result. (7) This and the rest of the entries will be made by the Quality Control Coordinator. Enter here (7) the name of the individual receiving the Report. (8) Enter the name of the individual initiating the Report. (9) The Quality Control Coordinator will check the entry in the laboratory notebook and note its presence. (10) The Quality Control Coordinator will note the necessity to initiate a Corrective Action Request (Figure 23-1) and attach a copy if the CAR is initiated. (11) The need for new samples will be indicated. (12) The fact that the customer was notified will be indicated. (13) If retesting or reanalysis of samples on hand is deemed necessary, it will be indicated here, and the name of the analyst or test technician to whom the work is assigned will be entered. (14) If an outside laboratory is used to confirm the results in question, the name and address of the laboratory will be entered. (15) The date that the Quality Control Coordinator completes the Review Report will be entered. (16) The Review Report will be signed by the Quality Control Coordinator.

22.3.3 The Quality Coordinator will take all steps necessary to prevent repetition of nonconformances including the initiation of a Corrective Action Request (Figure 23-1).q.v.

NONCONFORMANCE EVENT REVIEW REPORT

SAMPLE IDENT. NO. (1) _____   DATE (2) _____

SAMPLE DESCRIPTION (3) _____
_____

METHOD (4) _____
_____

SUSPECTED NONCONFORMANCE (5) _____
_____
_____

EXPECTED RESULT (6) _____

ACTIONS TAKEN

1. REPORTED TO: (7) _____

2. REPORTED BY: (8) _____

2. EVENT LOGGED IN LABORATORY NOTEBOOK: YES( ) NO( ) (9)

3. CORRECTIVE ACTION REQUEST INITIATED: YES( ) NO( ) (10)   (ATTACH COPY)

4. NEW SAMPLES REQUESTED: YES( ) NO( ) (11)

5. CUSTOMER NOTIFIED: YES( ) NO( ) (12)

6. RETEST OR REANALYSIS NECESSARY: YES( ) NO( ) (13)
   IF YES, TO WHOM ASSIGNED _____

7. THIRD PARTY LABORATORY CONFIRMATION NECESSARY: YES( ) NO( ) IF YES, NAME OF
   LABORATORY (14) _____

REVIEW DATE (15) _____   (16) _____

                                  QUALITY CONTROL COORDINATOR

FIGURE 22-1

23.1 Purpose

This section describes the procedures for correcting a departure from standard of any sort, fixing responsibility for the action to be taken, documenting the steps taken, and securing a report on the resolution of the problem.

23.2 Scope

This section deals with both short-term or immediate corrective action and long-term corrective action taken to avoid recurrence of the problem.

23.3 Kinds of Corrective Action

There are two types of corrective action available to the laboratory: on-the-spot or immediate, to correct or repair non-conforming data or equipment. Corrective action—long term, to eliminate causes of non-conformance and take measures to see that they do not recur. Steps comprising a closed-loop corrective action system are:

a. Define the problem.

b. Assign responsibility for investing the problem.

c. Determine a corrective action to eliminate the problem.

d. Assign and accept responsibility for implementing the corrective action.

e. Establish the effectiveness of the corrective action and implement the correction.

f. Verify that the corrective action has eliminated the problem.

Corrective action procedures recognize the need for a designated individual to test continually the effectiveness of the system and the corrective actions.

23.3.1 <u>On-the-spot or Immediate Corrective Action</u>—This is the process of correcting malfunctioning equipment.

The operator has the responsibility for conducting immediate corrective action when a problem arises.

Appropriate procedures are:

a. Found in the method being used, or-

b. Provided by the laboratory as a trouble-shooting checklist, or-

c. Provided by equipment manufacturers as trouble-shooting guides, or-

d. In the event that the solution to the problem is not found in the sources above, or is beyond the capabilities of the operator, he will report his findings to higher authority.

23.3.2 <u>Long-term Corrective Action</u> taken for recurring problems, will be initiated by the use of the Corrective Action Request form (Figure 23-1) which will be completed as follows: (1) Enter the Corrective Action Request (CAR) number. (2) Enter the suspense date. This is the date by which a solution is expected. (3) Enter the name and address of the originator of the CAR. (4) Enter the date that the CAR is originated. (5) Enter the name of the individual to whom the responsibility is given for arriving at a solution to the problem. (6) Enter the name of the organization of the person assigned the responsibility in (5). (7) Describe the nature of the problem in as much detail as necessary for a clear understanding of the problem. (8) If known, describe the underlying cause of the problem. (9) Enter the signature of the person originating the CAR. (10) Describe the phase of the operation or method during which the problem was discovered. (11) Enter the name and address of the person to whom the solution to the problem is to be reported, and the date by which the report is to be completed and returned. (12) List the names and addresses of other individuals to whom copies of the report are to be sent. (Reverse side–13) Repeat the CAR number at the top of the reverse side. (14) Enter the details of the recommended corrective action. (15) Indicate whether or not this is a temporary fix. (16) Indicate if this is a permanent solution. (17) To be signed by the person to whom the responsibility for completion of the corrective action was assigned. (18) Enter the date of the completed corrective action. (19) List the name of the individuals to whom copies of the completed CAR are to be routed. (20) Enter the initials of the authority approving the CAR. (21) Enter the date initialed. (22) Enter the suspense date that a follow-up is to be made to ensure that the recommended action has taken place. (23) Enter the initials of the authority approving the corrective action measures recommended.

When several CAR's are simultaneously being processed, it will be necessary to record them on the Corrective Action Master Log, (Figure 23-2) which will be filled out as follows: (1) Enter the assigned CAR No. from the first heading on the CAR. (CAR numbers will be carried across the top of this form. In this example there is room for five simultaneous entries.) (2) Enter the date submitted from Entry (4) of the CAR. (3) Enter the date logged in. (4) Enter the name of the originator of the CAR from Entry (3) of the CAR. (5) Enter a brief description of the problem presented. (6) Furnish a brief description of the cause of the problem, if known. (7) Enter the name of the individual to whom the problem was assigned for solution. (8) Indicate the suspense date. (9) Describe the nature or kind of investigation undertaken, if known. (10) Indicate the cause of the problem, when discovered. (11) Describe the corrective action recommended. (12) Indicate who is to be responsible for taking the recommended corrective action. (13) Indicate how the originator of the CAR is to be notified or was notified of the outcome of the investigation. (14) Indicate the results of any follow-up investigation conducted to determine whether the recommended action was taken and if it was effective in correcting the problem. (15) The Quality Control Coordinator will initial to indicate that the CAR was satisfactorily closed out.

Regardless of who initiates the CAR, The Quality Control Coordinator is responsible for the maintenance of the Corrective Action Request Master Log and for preparation of any periodic CAR status report required by management.

CORRECTIVE ACTION REQUEST

Corrective Action Request Form No. (1)   Suspense Date: (2)

Originator (3)   Date (4)

Person responsible for replying (5)   Organization (6)

Nature of problem below: (7)

P R O B L E M   I D E N T I F I C A T I O N

Cause of problem: (8)

Signature   (9)

During what phase of operation was the problem identified?   (10)

R E T U R N

TO: Name (11)
      Address
       By: (Date)

Information copies to: (Names and addresses) (12)

FIGURE 23-1

Corrective Action Request Form No. _____(13)_____

NOTE: Please return this form to the Originator with the corrective action planned on or before commitment date. If you determine that the action required is beyond the scope of your position, please indicate below who is responsible.

Corrective Action Planned. Please indicate effectivity point and dates.

C O R R E C T I V E   A C T I O N

_____(14)_____
_____
_____
_____
_____
_____
_____

Is this a temporary action? (15) Yes ( ) No ( )
Is this a permanent action? (16)   ( ) No ( )

Signature_____(17)_____

Date _____(18)_____

CAR
FINAL DISTRIBUTION
Original: Originator

(19)

ACCEPTANCE OF CA

Initially
Approved By: _____(20)_____

Date: _____(21)_____

Suspense Data
for Follow-up _____(22)_____

Final
Approved By: _____(23)_____

Date: _____(24)_____

THIS SPACE
RESERVED FOR
FOLLOW—UP

CORRECTIVE ACTION REQUEST
MASTER LOG

Corrective Action Request Form No.	(1)				
Date Submitted	(2)				
Date Received	(3)				
Submitted By	(4)				
General Problem Description	(5)				
Suspected Cause	(6)				
Assigned To	(7)				
Answer Due	(8)				
Nature of Problem Investigation	(9)				
Cause of Problem	(10)				
Corrective Action	(11)				
Corrective Action Responsibility	(12)				
Originator Notified	(13)				
Demonstration of Effectiveness	(14)				
Final Close-Out	(15)				

FIGURE 23-2

24.1 <u>Purpose</u>

This section describes how operating costs will be allocated according to quality assurance elements and grouped by cost categories and sub-categories. Quality control cost reporting is also required by this section.

24.2 <u>Scope</u>

This section deals only with operating costs related to the quality control function and does not consider other costs related to such functions as advertising, marketing, depreciation, utilities, taxes, etc.

24.3 <u>Reasons for Identifying Quality Costs</u>

Quality assurance costs should be identified and recorded primarily to identify elements whose costs are disproportionate to the benefits derived. An additional purpose is to detect cost trends for budget forecasting. Identification of costs is a prerequisite to cost reduction efforts.

24.4 <u>Definitions of Quality Cost Categories</u>

24.4.1 Prevention costs are those expenses incurred in keeping unacceptable data from being generated in the first place. Included in this category are such things as quality planning, quality training, etc.

24.4.2 Appraisal costs are those caused by the necessity to evaluate determinations or results that do not meet acceptance standards.

24.4.3 Internal failure costs are those caused by inadequate or spoiled samples, defective testing materials or reagents, malfunctioning instrumentation, or operator failures which lead to the production of unacceptable data.

24.4.4 External failure costs are those arising from incorrect or nonconforming data reported to customers. In other words, the defective product has already gone outside the laboratory.

24.5 Sub-categories of Laboratory Quality Costs—The elements of quality costs will be allocated to the four major cost categories as follows:

24.5.1 Prevention Costs

    a. Quality planning.

    b. Document control and revision.

    c. Quality training.

    d. Quality assurance plans for special projects and programs.

    e. The Quality Assurance Manual.

    f. Preventive maintenance.

24.5.2 <u>Appraisal Costs</u>

    a. Quality assurance activities associated with pre-test preparation, sample collection, sample analysis and test, and data reporting.

    b. Data validation.

    c. Statistical analysis of data.

    d. Procurement quality control.

    e. Calibration.

    f. Interlaboratory and intralaboratory testing.

    g. Audit procedures.

    h. Quality reports to management.

24.5.3 <u>Internal failure costs</u>

    a. Additional sample collection and additional analysis or testing required because of invalid samples.

    b. Reanalysis or retesting of collected samples because of equipment failure or operator error.

    c. Scrapping of data, because data was determined to be invalid.

24.5.4 <u>External failure costs</u>

    a. Investigation of customer complaints.

    b. Corrective action efforts.

    c. No charge repetition of tests or analysis to satisfy customer complaints.

24.6 <u>Distribution of quality costs</u>

After quality cost elements have been listed it will be necessary to allocate to them cost figures available from the accounting system. If actual costing figures are not available, the accounting component will provide estimates as nearly accurate as possible.

24.7 <u>Quality Cost Reports</u>

Quality control cost figures will be reported quarterly to management, showing cost figures allocated to each of the four major quality cost categories and the relationship of category costs to total quality costs.

The Quality Cost Report (Figure 24-1) will be prepared as a bar graph showing:

a. Expenditure by cost category.

b. Total quality cost expenditure.

c. Quarterly quality cost trends.

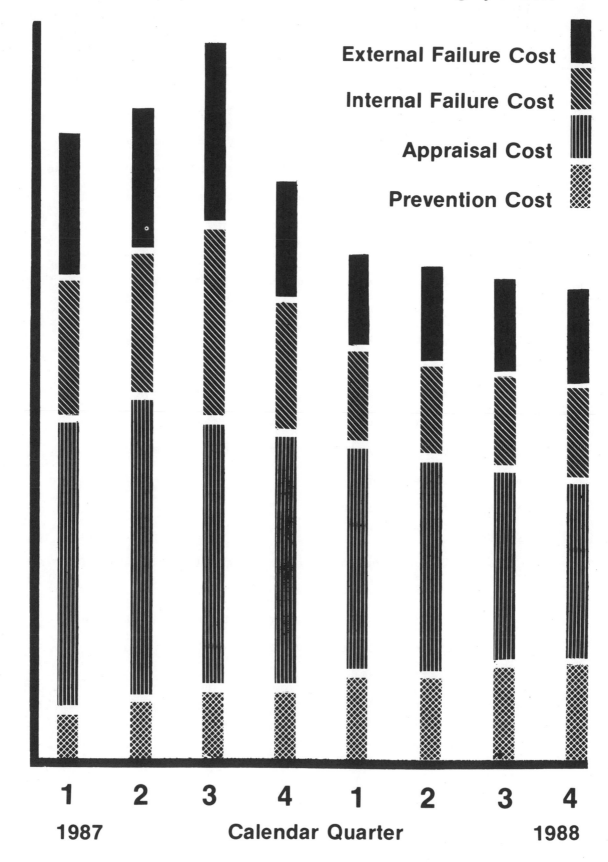

FIGURE 24-1

147

25.1 Purpose

This section describes the conduct of Quality System Audits. A check list is appended for use in carrying out system audits.

A system audit is an on-site, formal inspection and review of the laboratory Quality Control System, taking place on a periodic announced basis, to verify the effectiveness of the laboratory's quality program, as described in the several sections of this manual. The Quality System Audit is a Quality Assurance activity.

25.2 Scope

This procedure applies to the following internal functions; Receiving, Receiving Inspection (if applicable), Sample Storage, Chain-of-Custody procedures, Calibration, Preventive Maintenance, Data Validation, and other areas that affect the quality of laboratory output, as described in this manual. The quality auditing of supplier and subcontractor activities is covered by Sections 8.0 and 19.0.

25.3 Procedure

25.3.1 The Quality Control Coordinator will be responsible for arranging for the conduct of Quality System Audits. They will be carried out by an auditor from an outside agency, or by an auditor or audit team selected from outside the laboratory Quality Control organization.

25.3.2 The auditor, or audit team will use the Quality System Audit Check List (Figure 25-1) as a guide when conducting the audit and use this manual to establish the criteria for determining the degree of compliance with the requirements of this document.

25.3.3 The steps to be followed in conducting the system audit are:

 a. Notification of the dates and times of the planned audit to all individuals concerned.

 b. Hold pre-audit conference.

 c. Conduct of the audit.

 d. Hold post-audit conference, critique, and wrap-up.

 e. Follow-up to determine if deficiencies discovered during the audit have been corrected.

25.3.4 The Quality Control Coordinator will be responsible for initiating any Corrective Action Request (see Section 23.0) made necessary as a result of the audit.

25.3.5 The Quality System Audit Check List following (Figure 25-1) will be used to conduct the audit.

 The cover page of the Audit Check List will be filled out as follows: (1) Enter the name and title of the individual making the survey, or the name and title of the head of the audit team. (2) Enter the survey results from the scoring grid on the following page showing the total applicable points, the total survey points and the percentage grade. (3) Enter the date.

25.3.5.1 The audit will be graded using the instructions given on the first page of the check list.

25.3.5.2 All personnel whose activities have been audited will be given the grading results, however no pass/fail grade will be established as the primary purpose of the audit will be to identify areas of weakness and non-conformity with policies and procedures established in this Manual.

QUALITY CONTROL
QUALITY SYSTEM AUDIT CHECK LIST

INSTRUCTIONS

Laboratory compliance with each question will be indicated by placing 0, 1, 2, 3, or NA on the line opposite each question.

0. Required but not being done.
1. Below acceptance standards.
2. Adequate with minor departures from good practices.
3. Adequate in all respects.
NA. (Not Applicable) when it is obvious that a question would be inappropriate, due to type activity, lack of facilities, special methods or other valid reasons.

At the conclusion of the survey, each category of questions should be analyzed and summarized for acceptability. Space is provided in the index for entering these results. Remarks or comments by the survey team will be placed on the blank page opposite the question. The following method shall be used in summarizing each category: Total the survey points for all applicable questions. Divide this total by the total applicable questions. Multiplied by the weighting factor of 3, this computation will result in a percent acceptability figure.

$$\frac{\text{Total points for applicable questions rated}}{\text{Number of applicable questions X 3}} = \text{Percent Acceptability}$$

Example:
For category 1, we have a total of 10 applicable questions and a weighting factor of 3 per question (10 x 3)

$$\text{Therefore} \frac{28}{10 \times 3} = 93\% \text{ (Percent acceptability for category)}$$

SURVEY BY AND DATE

_____(1)_____ SURVEY RESULTS __(2)__ OF _____ = _____%

_____

_____

_____

NAME — TITLE — DATE (3)

FIGURE 25-1

QUALITY SYSTEM AUDIT CHECK LIST

INDEX

CATEGORY	SUBJECT	SURVEY RESULTS OF %
1	Quality Goals and Objectives	
2	Quality Policies	
3	The Quality Organization	
4	Management of the Quality Manual	
5	Quality Planning	
6	Quality in Procurement	
7	Sample Handling, Storage and Shipping	
8	Chain-of-custody Procedures	
9	Laboratory Testing and Analysis Control	
10	Quality Documentation and Records Control	
11	Control of Measuring and Test Equipment	
12	Preventive Maintainance	
13	Reference Standards	
14	Data Validation	
15	Environmental Control	
16	Customer Complaints	
17	Subcontracting	
18	Personnel Training, Qualification & Motivation	
19	Statistical Methods	
20	Nonconformity	
21	Corrective Action	
22	Quality Cost Reporting	
23	Quality Audits	
24	Reliability	

FIGURE 25-1

CATEGORY 1

QUALITY GOALS AND OBJECTIVES

1. Are the Quality Objectives clearly stated?   _____

2. Are the Quality Objectives quantified insofar as

    possible?   _____

3. Have the stated Quality Objectives been approved

    by management?   _____

                              Total Applicable Points _____

                              Total Survey Points   _____

                              Percentage   _____

CATEGORY 2

QUALITY POLICIES

1. Are the Quality Policies clearly stated?   _____

2. Do the Quality Policies support the

    achievement of quality objectives previously

    stated?   _____

3. Are the Quality Policies approved by management? _____

                              Total Applicable Points_____

                              Total Survey Points   _____

                              Percentage   _____

FIGURE 25-:1

CATEGORY 3

THE QUALITY ORGANIZATION

1. Laboratory management reflects a positive
   attitude toward Quality Control   _____

2. Direct responsibility for Quality Control has been
   formally established.   _____

3. The functional relationship between Quality
   Control and other laboratory components has been
   clearly delineated.   _____

4. The line of authority within the Quality Control
   organization has been clearly delineated.   _____

5. The job description for the Quality Control
   Coordinator is up-to-date and accurately describes
   the current duties, responsibilites and authority of
   that individual.   _____

6. The Quality Control Coordinator is not subordinate
   to any individual who is responsible for carrying
   out testing or analytical responsibilities.   _____

                            Total Applicable Points _____
                            Total Survey Points   _____
                            Percentage   _____

FIGURE 25-1

154

CATEGORY 4

MANAGEMENT OF THE QUALITY MANUAL

1. The responsibility for maintenance of the Quality

   Manual is clearly established. _____

2. The issue of the manual is controlled. _____

3. Provision is made for updating the Manual and

   issuing revisions to holders of controlled

   copies. _____

4. Provision is made for the issue of uncontrolled

   copies as necessary. _____

                              Total Applicable Points _____

                              Total Survey Points _____

                              Percentage _____

CATEGORY 5

QUALITY PLANNING

1. The responsibility for the quality planning

   function is clearly defined. _____

2. There is objective evidence that quality

   planning is an on-going, continuous process. _____

3. The Quality Manual accurately reflects current

   quality system activities in the laboratory. _____

                              Total Applicable Points_____

                              Total Survey Points _____

                              Percentage _____

FIGURE 25-1

CATEGORY 6

QUALITY IN PROCUREMENT

1. Inspection is performed by Quality Control. _____

2. Inspection instructions are available to
    inspection personnel. _____

3. Inspection personnel perform inspection operations
    in accordance with current instructions and
    procedures. _____

4. There is a system to maintain inspection
    instructions current. _____

5. There is a program for Quality Control approval
    of vendors and subcontractors. _____

6. Vendors and subcontractors are required to have a
    a system for quality control. _____

7. Inspection and test results are used to evaluate
    vendor quality performance. _____

8. Purchase orders, specifications, and vendors'
    catalogs are available to receiving inspection. _____

9. Inspection gages and test equipment are adequate to
    perform required inspection. _____

10. Sampling inspection (when applicable) is performed
    to ANSI/ASQC Z 1.4-198/, ANSI/ASQC Z 1.9-1980,
    or other approved sampling plan. _____

11. Approved laboratories are used for special
    tests and analyses outside the scope of the
    laboratory's capabilities. _____

FIGURE 25-1

12. All reagent stock is properly labeled, dated and
marked with shelf-life expiration when necessary. _____

13. Chemical/physical analysis, test specifications,
or certifications are required when applicable. _____

Total Applicable Points _____

Total Survey Points _____

Percentage _____

CATEGORY 7

SAMPLE HANDLING, STORAGE AND SHIPMENT

1. The Quality Control Coordinator monitors the
physical condition of incoming test samples. _____

2. Test and analytical samples are properly
identified. _____

3. Sample Submittal Forms are completely and
accurately filled out. _____

Total Applicable Points _____

Total Survey Points _____

Percentage _____

CATEGORY 8

CHAIN-OF-CUSTODY PROCEDURES

1. Is the Chain-of-Custody Documentation Form correctly
and completely filled out? _____

2. Does the Chain-of-Custody Documentation Form
establish custody of the sample at all times? _____

Total Applicable Points _____

Total Survey Points _____

Percentage _____

FIGURE 25-1

CATEGORY 9

LABORATORY TESTING AND ANALYSIS CONTROL

1. Function checks are routinely performed by the operator. _____

2. The operator performs control checks during the analytical or testing procedures. _____

3. Tests for significance of difference are made when results indicate the need. _____

4. Control charts are used routinely for suspected or difficult method determinations, _____

5. The Quality Control Coordinator is responsible for deciding which testing or analytical operations will be charted. _____

Total Applicable Points_____

Total Survey Points _____

Percentage _____

CATEGORY 10

QUALITY DOCUMENTATION AND RECORDS CONTROL

1. The quality documents to be controlled are listed in the Quality Manual. _____

2. Document retention policies are published in in the Quality Manual or elsewhere in writing. _____

3. The Quality System is adequately documented. _____

4. The Quality Control Coordinator issues periodic reports to management which reflect quality levels, quality costs, customer complaints, etc. _____

FIGURE 25-1

5. There is adequate control over the issue of new
technical documents and changes. _____

6. Obsolete documents are retrieved from work stations
and are not permitted to remain in the hands of
users. _____

Total Applicable Points _____

Total Survey Points _____

Percentage _____

CATEGORY 11

CONTROL OF MEASURING AND TEST EQUIPMENT

1. Quality Control procedures insure that instruments,
gages, measuring devices and test equipment are
periodically inspected and recalibrated at
established intervals. _____

2. Working and reference standards of required
accuracy that are periodically calibrated to
primary standards traceable to the NIST are
maintained in a suitable environment. _____

3. New or repaired test equipment and measuring
devices are calibrated and proved satisfactory
by Quality Control before release for use. _____

4. There are written calibration procedures for each
type of instrument and measuring device. _____

5. Objective evidence includes calibration date and
date next calibration must be performed on
measuring instrumentation. _____

FIGURE 25-1

6. Record system provides for automatic recall of gages, instruments and test equipment due for calibration. _____

                                        Total Applicable Points _____

                                        Total Survey Points _____

                                        Percentage _____

CATEGORY 12

PREVENTIVE MAINTENANCE

1. There is a planned, scheduled program tied in with the calibration system, that ensures that all measuring and test equipment undergoes periodic preventive maintenance. _____

2. ·Preventive maintenance checklists are in use and on file. _____

3. An adequate spare parts inventory is on hand. _____

                                        Total Applicable Points _____

                                        Total Survey Points _____

                                        Percentage _____

FIGURE 25-1

CATEGORY 13

REFERENCE STANDARDS

1. Laboratory policy requires the use of reference

    standards, when available,   _____

2. The Quality Control Coordinator keeps on file

    a copy of NIST Publication 260 and copies of

    NIST Certicates of Analysis.   _____

                           Total Applicable Points

                           Total Survey Points   _____

                           Percentage   _____

CATEGORY 14

DATA VALIDATION

1. Data validation is routinely employed to detect

    outliers or spurious values.   _____

2. Manual data validation techniques are the only

    methods used to detect outliers and spurious values. _____

3. Both manual and computerized techniques are used

    for data validation.   _____

4. Sampling plans from ANSI/ASQC Z1.4, 1981 are used

    when selecting samples for data validation of large

    quantities of data.   _____

                           Total Applicable Points _____

                           Total Survey Points   _____

                           Percentage   _____

FIGURE 25- 1

CATEGORY 15

ENVIRONMENTAL CONTROLS

1. The laboratory working environment is controlled

    to the extent prescribed by the Manual.   _____

                              Total Applicable Points _____

                              Total Survey Points   _____

                              Percentage   _____

CATEGORY 16

CUSTOMER COMPLAINTS

1. One individual is charged with the responsibility

    for handling all technical complaints from

    customers, regulatory agencies, and accreditation

    organizations.   _____

2. There is a system for providing feedback to all

    individuals concerned.   _____

3. One individual is charged with the responsibility

    for initiating corrective action if this becomes

    necessary.   _____

                              Total Applicable points

                              Total Survey Points   _____

                              Percentage   _____

FIGURE 25-1

162

CATEGORY 17

SUBCONTRACTING

1. Contract laboratories are required to have a viable Quality Control System in place. _____

2. Provision is made for furnishing contract laboratories with "audit" samples to test laboratory proficiency. _____

3. When a new method is introduced, provision is made for furnishing contract laboratories with spiked samples to test the proficiency of the laboratory in performing the method. _____

               Total Applicable Points _____

               Total Survey Points _____

               Percentage _____

CATEGORY 18

PERSONNEL TRAINING, QUALIFICATION AND MOTIVATION

1. The laboratory provides for employee indoctrination and training in Quality Control _____

2. The laboratory has a program to test the effectiveness of the training effort. _____

3. Training records are available to indicate the status of individual training. _____

4. The laboratory conducts campaigns, supported by management, to promote the achievement of superior quality output. _____

               Total Applicable Points _____

               Total Survey Points _____

               Percentage _____

FIGURE 25-1

CATEGORY 19

STATISTICAL METHODS

1. The laboratory employs the following statistical
    techniques in the routine conduct of analytical
    and testing work:

          a.   Tests of hypotheses   _____

          b.   Treatment of outliers   _____

          c.   Control Charts   _____

          d.   Sampling Plans   _____

          e.   Tests for significance of differences   _____

                              Total Applicable Points_____

                              Total Survey Points   _____

                              Percentage   _____

CATEGORY 20

NONCONFORMITY

1. The Quality Control System assures that positive
    action is taken, when a nonconforming event occurs,
    to ensure that all required tasks are performed in
    accordance with provisions of the Quality Manual.   _____

2. Adequate records are kept of nonconformities.   _____

                              Total Applicable Points _____

                              Total Survey Points   _____

                              Percentage   _____

FIGURE 25-1

CATEGORY 21

CORRECTIVE ACTION

1. The laboratory has a method of recurrence control
    to prevent repetitive discrepancies.   _____

2. Corrective action requests on nonconforming events
    are initiated promptly.   _____

3. Records provide for follow-up on all corrective
    action requests.   _____

4. The responsibilty for solving the problem by
    taking corrective action is clearly established.   _____

5. The Suspense Date when the corrective action is
    expected to be completed is clearly established.   _____

6. The Corrective Action Log controls the progress
    of the corrective action initiated by the
    Corrective Action Request.   _____

                                Total Applicable Points _____

                                Total Survey Points   _____

                                Percentage   _____

FIGURE 25-1

CATEGORY 23

QUALITY COST REPORTING

1. Quality costs are accumulated for the following

   categories:

      a. Prevention costs   _____

      b. Appraisal costs   _____

      c. Internal failure costs   _____

      d. External failure costs   _____

2. When actual costs figures cannot be obtained from

   the cost accounting component, reasonably

   accurate estimates are used.   _____

3. Quality Cost reports are made to management

   periodically.   _____

4. Quality cost figures are used in the annual

   budgeting process.   _____

               Total Applicable Points   _____

               Total Survey Points   _____

               Percentage   _____

FIGURE 25-1

CATEGORY 24

QUALITY AUDITS

1. Quality audits are performed on and announced,
   scheduled basis by individuals outside the
   quality organization.   _____

2. All areas of the laboratory contributing to
   the quality of work output are audited.   _____

3. The results of the audit are reported to
   management in writing.   _____

4. Pre-audit and post-audit conferences are held
   prior to and after each audit.   _____

                                    Total Applicable Points   _____

                                    Total Survey Points   _____

                                    Percentage   _____

CATEGORY 25

RELIABILITY

1. Reliability forecasts in terms of "Mean Time
   to Failure" (MTTF) are routinely provided by Quality
   Control for remotely situated recording
   instruments,   _____

                                    Total Applicable Points   _____

                                    Total Survey Points   _____

                                    Percentage   _____

FIGURE 25-1

26.1 <u>Purpose</u>

This section describes the responsibilities of the Quality Control Coordinator with regard to ensuring the reliability of remote recording instruments used by the laboratory.

26.2 <u>Scope</u>

The actions required in this section apply only to measurement equipment located off-site, in remote areas where there is no day-to-day oversight on equipment operation. This section is not applicable to gages, instruments, and equipment used within the laboratory.

26.3 <u>Actions and Responsibilities-Quality Control Coordinator</u>

In order to ensure high reliability (completeness of data) the Quality Control Coordinator is responsible for the following actions:

a. He must see that equipment reliability is specified in purchase contracts.

b. Incoming equipment must be inspected and tested for adherence to contract specifications by burn-in or demonstrated performance acceptance testing.

c. The operating environment which influences the reliability of measurements must be controlled insofar as possible.

d. Provison must be made for adequate training of personnel.

e. Ensure that periodic preventive maintenance is carried out to minimize wear-out failures. (Section 14.0)

f. Keep records of failures, and analyze and use these data to provide the basis for the initiation of corrective actions and predict reliability rates.

26.4 The technician operating the equipment has the following responsibilities to ensure high reliability (completeness of data):

a. When the measurement system is operational, routine preventive maintenance will be performed.

b. When the system fails to operate as required, the time of failure will be noted on the equipment strip chart or in the laboratory notebook, whichever is applicable. In addition, the top portion (blocks 1-13) of the Maintenance Service Report (Figure 26-1) will be filled out as follows: (1) Enter the name of the equipment. (2) Enter the identification number of the test being run on the failed equipment. (3) Enter the date and time the test was begun. (4) Give the name of the equipment manufacturer. (5) Enter the time the failure was noted. (6) Enter the date the failure was discovered. (7) Give the model number of the failed piece of equipment. (8) Give the operator's name. (9) Identify the location of the equipment. (10) Enter the interval since the last recorded maintenance or repair service. (11) Give a general description of the nature of the equipment failure, and its cause, if known, (12) Describe the environmental conditions obtaining at the time of failure as closely as possible. (13) Enter the time Md, spent on diagnosing the failure, and its cause.

c. After the measurement system has been repaired, complete the Maintenance Service Report as follows: (14) Describe, in detail the action taken to repair the equipment. (15) List the failed components replaced. (16) List the failed components repaired. (17) Note whether tamperproof seals were required. (18) Describe what preventive maintenance work was completed. (19) Note what operational tests were conducted and completed. (20) Record Time, Mr for total repair, preventive maintenance, and operational testing. (21) The individual completing the Report will sign at the bottom where indicated.

d. The operator will deliver to the Quality Control Coordinator the original of the Maintenance Service Report for calculation of Uptime U, Downtime D, and Availability.

$$A = U/D + U$$

MAINTENANCE / SERVICE REPORT

ITEM NAME (1)	TEST NO. (2)	TIME STARTED (3)
MANUFACTURER (4)	TIME (5)	DATE (6)
MODEL (7)	OPERATOR (8)	

FAILURE CONDITIONS

LOCATION (9)	INTERVAL SINCE LAST SERVICE (10)

GENERAL DESCRIPTION OF FAILURE AND CAUSE
(11)

ENVIRONMENTAL CONDITIONS (12)	DIAGNOSTIC TIME SPENT (13)

CORRECTIVE ACTION TAKEN
(14)

FAILED COMPONENTS REPLACED
(15)

FAILED COMPONENTS REPAIRED
(16)

SERVICE REQUIREMENTS

TAMPER—PROOF SEALING REQUIRED ☐ (17)   NOT REQUIRED ☐

PREVENTIVE MAINTAINANCE COMPLETED (18)

OPERATIONAL TEST COMPLETED (19)

_____
SIGNATURE

FIGURE 26-1

	Effective Date	Subject	Section No. 27.0
	By	QUALITY MANUAL FORMAT SHEET	Page 1 of 2
	Approved		Revised

27.1 <u>Purpose</u>

This section provides instruction in how to use the Manual standard format sheet. (Figure 27-1)

27.2 <u>Scope</u>

These instructions apply to all those individuals responsible for the preparation of the Quality Manual or its revisions.

27.3 <u>Use of the Manual Standard Format Sheet</u>

The Laboratory Quality Manual Standard Format Sheet will be typewritten or prepared on a word processor, and will be filled out as follows: (1) This space may be used to show the laboratory logotype. (2) Enter the date that this page becomes effective and the old document becomes obsolete. (3) Enter the name of the author or person preparing this document. (4) Enter the name or initials of the individual approving this document. (5) Enter the title of the quality element being addressed. (6) Indicate the Section Number of the element, taken from the Table of Contents. (7) Enter the page number of the document. Pages in each section will be numbered consecutively and will begin with one (1) at the start of each new section. (8) In the case of revised documents, enter the revision date. (9) Type the text material in the space provided.

(1)	Effective Date (2)	Subject	Section No. (6)
	By (3)	(5)	Page (7) of
	Approved (4)		Revised (8)

(9)

FIGURE 27-1

Part 4

SAMPLE QUALITY ASSURANCE FORMS

On the following pages are examples of forms used in the laboratory quality system. They are presented full-sized so that they can be reproduced. Of course, they may be changed in any way that makes them applicable to the individual laboratory's procedures and work methods. The forms are the same as those shown in the body of the sample Laboratory Quality Assurance Manual in Part 3, with the instructional reference numbers removed. The figure numbers are identical.

SAMPLE QUALITY ASSURANCE FORMS

CONTENTS

XYZ LABORATORIES, INC.

RECEIVING REPORT

Date _____ 19 _____

Received From _____

Shipped From. _____

Shipped Via

SHIPPING CHARGE	PREPAID	COLLECT	FREIGHT BILL NO	CAR INITIALS AND NO

PURCHASE ORDER NO		FOR DEPT	

QUANTITY	DESCRIPTION OF MATERIALS	CONDITION

Remarks: _____

_____

_____

_____

_____

RECEIVING CLERK

FIGURE 8-1

XYZ LABORATORIES, INC.
QUALITY ASSURANCE
CLASSIFICATION OF DEFECTS
INSPECTION INSTRUCTIONS

DESCRIPTION _____

PART NUMBER _____ LOT SIZE _____

SAMPLING PROCEDURE: ANSI / ASQC / Z 1.4 Level II, unless otherwise noted.

CRITICAL
100% (no defects allowed)

INSPECTION PROCEDURE   **METHOD**

1. _____ | _____

2. _____ | _____

3. _____ | _____

4. _____ | _____

5. _____ | _____

6. _____ | _____

MAJOR "A"
AQL 1.0 %

101. _____ | _____

102. _____ | _____

103. _____ | _____

104. _____ | _____

105. _____ | _____

106. _____ | _____

MAJOR "B"
AQL 2.5 %

201. _____ | _____

202. _____ | _____

203. _____ | _____

204. _____ | _____

205. _____ | _____

MINOR
AQL 4 %

301. _____ | _____

302. _____ | _____

303. _____ | _____

304. _____ | _____

305. _____ | _____

FIGURE 8-2

TESTING AND ANALYTICAL SUPPLIES

RECEIVING AND STORES LOG

Log No.	Identification	P.O. No.	Vendor	Date	Disposition Acc. - Rej.

FIGURE 8-3

XYZ LABORATORIES, INC.

QUALITY SYSTEM SURVEY EVALUATION CHECK LIST

COMPANY _____

ADDRESS_____ PHONE_____

GENERAL PRODUCTS _____

TOTAL NUMBER OF EMPLOYEES _____ ,QC_____

PERSONNEL CONTACTED AND PRODUCTS CONSIDERED

SURVEY BY AND DATE

_____   SURVEY RESULTS _____ OF _____=_____ %

_____

_____

_____
     NAME — TITLE — DATE

FIGURE 8-4

INDEX

INSTRUCTIONS

An applicant's compliance to each question shall be indicated by placing 0, 1, 2, 3, or NA on the line opposite each question.

- 0. Required but not being done.
- 1. Below acceptance standards.
- 2. Adequate with minor departures from good practices.
- 3. Adequate in all respects.
- NA. (Not Applicable) when it is obvious that a question would be inappropriate, due to type of product, lack of facilities, special processes or other valid reasons.

At the conclusion of the survey, each category of questions should be analyzed and summarized for acceptability. Space is provided in the index for entering these results. Remarks or comments by the survey team will be placed on the blank page opposite the question.

The following method shall be used in summarizing each category: Total the survey points for all applicable questions. Divide this total by the total applicable questions. Multiplied by the weighting factor of 3, this computation will result in a percent acceptability figure.

$$\frac{\text{Total points for applicable questions rated}}{\text{Number of applicable questions X 3}} = \text{Percent Acceptability}$$

Example:

For category 1, we have a total of 10 applicable questions and a weighting factor of 3 per question (10 X 3)

Therefore $\frac{28}{10 \times 3} = 93\%$ (Percent acceptability for category)

————————————————

CATEGORY 1
QUALITY CONTROL ORGANIZATION

1. Company management reflects a positive attitude toward Quality Control. _____
2. Direct responsibility for Quality Control has been formally established. _____
3. The functional relationship between Quality Control and other departments has been clearly delineated. _____
4. Line of authority within the Quality Control organization has been clearly delineated. _____
5. The Quality Control System has been designed to promote defect prevention. _____
6. The Quality Control System is adequately documented. _____
7. The Quality Control Department issues periodic reports to management which reflect quality levels, rework, scrap, work realization, quality costs, etc. _____
8. The documented Quality Control System carries management approval. _____
9. The Quality Control Department adequately provides for the use and control of inspection stamps. _____
10. The applicant provides for employee indoctrination and training in Quality Control methods. _____
11. Quality Control System assures that Engineering drawings and specification requirements, procured materials and manufactured items satisfy all customer contractual requirements. _____

Total Applicable Points _____
Total Survey Points _____
Percentage of Acceptability _____

- 1 -

CATEGORY 2
CONTROL OF PROCURED SUPPLIES

1. <u>OPERATION AND DOCUMENTATION</u>
   a. Inspection is preformed by Quality Control. _____
   b. Inspection instruction/procedures are available to inspection personnel. _____
   c. Inspection personnel perform inspection operations in accordance with current instructions and procedures. _____
   d. Applicant has provided a system to maintain instruction/procedures current. _____

2. <u>VENDOR PROGRAM</u>
   a. Applicant's Quality Control reviews environmental and life requirements and assures that Qualified Products are being procured. _____
   b. Applicant has an established program for processing drawing changes to his vendors and subcontractors. _____
   c. Receiving Inspection records reflect quality history of applicant's venuors and subcontractors. _____
   d. Vendor quality performance data reports are used by other departments in making procurement decisions. _____
   e. Applicant has program for Quality Control approval of vendors and subcontractors. _____
   f. Applicant performs source surveillance at subcontractor's plant when applicable. _____
   g. Applicant requires his vendors and subcontractors to have a system for quality control. _____
   h. Applicant uses inspection/test results to contribute to vendors evaluation. _____

3. <u>INSPECTION CRITERIA</u>
   Purchase Orders, Drawings, Engineering Orders, specifications, and vendor's catalogs are available to Receiving Inspection. _____

4. <u>INSTRUMENTATION</u>
   a. Inspection gages and test equipment are adequate to perform required inspection. _____
   b. Tools, fixtures, and inspection equipment are identified, stored and issued under controlled conditions. _____

5. <u>SAMPLING INSPECTION</u> (When Applicable)
   a. Sampling Inspection is performed to MIL-STD-105, MIL-STD-414, or other approved plan. _____
   b. Inspection personnel are provided with instructions covering Sampling Inspection. _____

6. <u>CONTRACTED SPECIAL PROCESSES</u>
   a. Applicant has established certified and/or approved source for special process services. _____
   b. Applicant's Quality Control performs surveillance over special processing. _____

7. <u>STOCK CONTROL</u>
   a. Applicant uses a positive means of identification on all stock. _____
   b. Applicant requires and receives chemical/physical analysis or test specifications when applicable _____
   c. Applicant provides adequate control area for customer furnished material when required. _____
   d. Applicant has an acceptable system for "age control" for items where acceptability is limited by maximum age. _____

Total Applicable Points _____
Total Survey Points _____
Percentage of Acceptability _____

- 2 -

185

CATEGORY 3
IN-PROCESS INSPECTION

1. **OPERATION AND DOCUMENTATION**
   a. In-process inspection is performed by Quality Control. _____
   b. Inspection instructions/procedures are available to Inspection personnel. _____
   c. Inspection personnel perform inspection operations in accordance with current instruction and procedures. _____
   d. Applicant has provided a system for First Article Inspection and re-inspection after changes in manufacturing process. _____
   e. Applicant procedures provide for inspection control of in-process material. _____
2. **INSPECTION CRITERIA**
   a. Illegible or obsolete drawings are not in use by inspection. _____
   b. Current Drawings, Engineering Orders, and specifications are available at Inspection area. _____
3. **INSTRUMENTATION**
   a. Adequate in-process inspection facilities are available. _____
   b. Inspection gages and test equipment are adequate to perform required inspection. _____
   c. Adequate surveillance and maintenance of inspection equipment when in use, transported or stored is maintained. _____
4. **STATISTICAL QUALITY CONTROL (When Applicable)**
   a. Sampling inspection is preformed to MIL-STD-105. MIL-STD-414, or other approved plan. _____
   b. Inspection personnel are provided with instructions covering sampling inspection. _____
   c. Control charts or other in-process statistical quality control methods are used. _____

Total Applicable Points _____
Total Survey Points _____
Percentage of Acceptance _____

CATEGORY 4
FINAL INSPECTION

1. **OPERATION AND DOCUMENTATION**
   a. Final inspection is performed by Quailty Control. _____
   b. Current Inspection instruction/procedures are available to Inspection personnel. _____
   c. Inspection personnel perform inspection operations in accordance with current instructions and procedures. _____
   d. Completed supplies are inspected as necessary to assure that contract requirements have been met. _____
   e. Environmental and life tests are periodically performed to assure that manufacturing and process degradation has not significantly effected design integrity. _____
2. **INSPECTION CRITERIA**
   Current Drawings, Engineering Orders, specifications and/or customer requirements are available at final inspection. _____
3. **INSTRUMENTATION**
   a. Inspection gages and test equipment are adequate to perform required inspection. _____
   b. Tools, fixtures, and inspection equipment are identified, stored, and issued under controlled conditions. _____
   c. Adequate final inspection facilities are available. _____
4. **STATISTICAL QUALITY CONTROL (When Applicable)**
   a. Sampling Inspection is performed to MIL-STD-105, MIL-STD-414, or other approved plan. _____

-3-

CATEGORY 4
FINAL INSPECTION
(Cont.)

    b.   Inspection personnel are provided with instructions covering sampling inspection. _____

    c.   Control charts or other statistical quality control methods are used. _____

5.   **MATERIAL HANDLING**

    Provisions are made to prevent unauthorized use of uninspected and/or conforming material. _____

                           Total Applicable Points   _____

                            Total Survey Points   _____

                            Percentage of Acceptability   _____

CATEGORY 5
SHIPPING INSPECTION

1.   **OPERATION AND DOCUMENTATION**

    a.   Shipping Inspection is under the surveillance of Quality Control. _____

    b.   Inspection operations are performed in accordance with current instructions and procedures. _____

2.   **INSPECTION CRITERIA**

    a.   Inspectors have packaging requirements. _____

    b.   Packaging tests are performed/witnessed as required by applicable specifications. _____

    c.   Certified packaging materials are used where required. _____

                            Total Applicable Points   _____

                            Total Survey Points   _____

                            Percentage of Acceptability   _____

CATEGORY 6
CALIBRATION OF MECHANICAL MEASURING EQUIPMENT

1.   Quality Control procedures insure that inspection gages, measuring devices, and test equipment are periodically inspected and recalibrated at established intervals. _____

2.   Production tooling which is in use as a medium of inspection is periodically inspected by Quality Control at intervals that assure the maintenance of required accuracy. _____

3.   Applicant has system for modification of inspection/test equipment to latest engineering changes. _____

4.   Applicant does maintain, in a suitable environment, working standards of required accuracy that are periodically calibrated to primary standards traceable to the National Bureau of Standards. _____

5.   Personally owned tools and gages show evidence of periodic recalibration. _____

6.   Specialized test equipment used for acceptance purposes is proved satisfactory by Quality Control before release for use. _____

7.   New or reworked test/inspection equipment is inspected and calibrated prior to use. _____

8.   Applicant has written instruction/procedures for "gage and test calibration." _____

9.   Objective evidence includes calibration date and date next calibration must be performed on measuring instrumentation. _____

10. Records are maintained for periodic recalibration of inspection gages and test equipment. _____

11. Record system provides for automatic recall of inspection gages and test equipment. _____

                            Total Applicable Points   _____

                            Total Survey Points   _____

                            Percentage of Acceptability   _____

- 4 -

187

CATEGORY 7
CALIBRATION OF ALL OTHER TEST EQUIPMENT

1. Quality Control procedures insure that inspection equipment is periodically inspected and recalibrated at established intervals. _____
2. Production tooling which is in use as a medium of inspection is periodically inspected by Quality Control at intervals that assure the maintenance of required accuracy. _____
3. Applicant has system for modification of inspection/test equipment to latest engineering changes. _____
4. Applicant does maintain, in a suitable environment, working standards of required accuracy that are periodically calibrated to primary standards traceable to the National Bureau of Standards. _____
5. Personally owned tools show evidence of periodic recalibration.
6. Specialized test equipment used for acceptance purposes is "proved" by Quality Control before release for use. _____
7. New or reworked test/inspection equipment is inspected and calibrated prior to use. _____
8. Applicant has written instruction/procedures for all Measuring Equipment Calibration. _____
9. Objective evidence includes calibration date and date next calibration must be performed on measuring instrument. _____
10. Records are maintained for periodic recalibration of all inspection measuring equipment. _____
11. Record system provides for automatic recall of all inspection measuring equipment. _____

Total Applicant Points _____
Total Survey Points _____
Percentage of Acceptability _____

CATEGORY 8
DRAWING AND CHANGE CONTROL

1. The direct and specific responsibility to verify that changes are incorporated at effective points is clearly defined and formally established. _____
2. Applicant has written procedures describing drawing change control. _____
3. A control system exists for the issue and return of each drawing. _____
4. Drawing and change control system prevents the use of illegible or obsolete drawings by Inspection. _____

Total Applicable Points _____
Total Survey Points _____
Percentage of Acceptability _____

CATEGORY 9
ENVIRONMENTAL TEST FACILITIES
(When Applicable)

1. Applicant has suitable environmental test equipment to perform full range of required tests on product under consideration, or uses recognized test laboratories. _____
2. Applicant maintains records that show results of tests performed and failure rates. _____
3. Applicant has established program of calibration and maintenance of instrumentation used in environmental laboratory. _____

Total Applicable Points _____
Total Survey Points _____
Percentage of Acceptability _____

- 5 -

188

CATEGORY 10
NON-CONFORMING MATERIAL
(Materials Review)

1. Applicant's procedures reflects control of authority to repair non-conforming supplies. _____
2. Applicant has a system for the diversion of non-conforming supplies from normal production flow. _____
3. Applicant has method of recurrence control to prevent repetitive discrepancies. _____
4. Corrective action requests on non-conforming materials are initiated promptly. _____
5. Records provide for follow-up on all corrective action requests. _____
6. Disposition of non-conforming material does not establish criteria for like dispositions. _____
7. Reports on non-conforming materials are issued to management for action. _____

Total Applicable Points _____
Total Survey Points _____
Percentage of Acceptability _____

CATEGORY 11
HOUSEKEEPING, STORAGE & HANDLING

1. Materials, supplies, and work in process are arranged in a neat workmanlike manner. _____
2. Work and storage areas are clean and free from dirt, rejected materials, and other materials which could contaminate or damage acceptable materials. _____
3. In the case of extended or indefinite storage, equipment is given proper preservative treatment. _____
4. Facilities for storage of materials or equipment assures prevention of deterioration and contamination. _____
5. During storage, material is packaged in a manner to prevent deterioration and damage. _____
6. Delicate instruments are handled in a manner that will not jeopardize the reliability of the design characteristics. _____
7. Transfer procedures are adequate to prevent comingling, contamination, and loss. _____
8. Tags and labels are properly used to indicate the identity, condition, and status of the equipment. _____
9. Components, assemblies, and materials are handled so as to prevent damage and deterioration. _____
10. Provisions are made to prevent unauthorized use of uninspected materials. _____

Total Applicable Points _____
Total Survey Points _____
Percentage of Acceptability _____

189

SAMPLE SUBMITTAL FORM

Sample Log No. _____ Date _____

Originator: _____

Address _____

_____

Telephone No. _____ Project No. _____

Sample Description. _____

Sampling Source. _____

Number of Samples Submitted. _____

Date of Collection _____

Date of Sample Shipment _____

Request for Analysis

Sample Field Number	Sample Characteristics			Test or Analyses Requested	Remarks
	Type	Manuf.	Lot No.		

Comments _____

_____

Possible Interfering Compounds_____

_____

FIGURE 9.1

190

CHAIN OF CUSTODY DOCUMENTATION

NAME OF SUBMITTER

ADDRESS:

DESCRIPTION OF SAMPLE:

SAMPLE SOURCE:

DATE AND TIME OF COLLECTION:

METHOD OF SHIPMENT:

_____

SAMPLE #:

TELEPHONE NUMBER: _____

DATE RECEIVED IN LAB: _____ RECEIVED BY WHOM: _____

RECEIPT CONDITION (CONTAINERS, PACKAGING, AND LABELING):_____

INITIAL WEIGHT OF CONTAINER (BULKS): _____

WHERE INITIALLY STORED:_____

ACCOUNTABILITY RECORD

REMOVAL DATE &TIME	BY WHOM	LOCATION	RETURNED AMOUNT (% OF INITIAL)	DATE TIME	TO WHOM	LOCATION

SAMPLE DISPOSITION:

DATE DISPOSED:

BY WHOM:

HOW DISPOSED:

FIGURE 10-1

XYZ LABORATORIES, INC.

TECHNICAL DOCUMENT CHANGE NOTICE

DATE _____

TDCN No. _____

EFFECTIVE _____

DOCUMENT TYPE
- _____ METHOD
- _____ SAMPLING DATA SHEET
- _____ CALIBRATION INSTRUCTION
- _____ OTHER (DESCRIBE)

DOCUMENT TITLE _____

REQUESTED BY: _____

CHANGE FROM: _____

_____

_____

_____

_____

CHANGE TO: _____

_____

_____

REASON FOR CHANGE: _____

_____

_____

DISTRIBUTIONS	SIGNATURES	COSTS	$	APPROVALS:
		NEW EQUIP.		
		NEW MAT'L		
		SCRAPPED OR OBS. EQUIP.		
		SCRAPPED OR OBS. MAT'L		
		PERSONNEL		
		OTHER		

DETAIL ON REVERSE SIDE

FIGURE 12-1

CALIBRATION FREQUENCY SCHEDULE

Type	Minimum Frequency	Instruction Number or Calibration Method

FIGURE 13-1

INSTRUMENT / GAUGE CALIBRATION RECORD

INSTRUMENT No.	DESCRIPTION	IDENTIFICATION No.

MFR.	SERIAL No.	MODEL No.

LOCATION	CALIBRATION FREQUENCY: EVERY_____	CALIBRATION PROCEDURE No.

DATE CALIBRATED	CHECKED BY	DEPT.	RESULTS

(OVER)

DATE CALIBRATED	CHECKED BY	DEPT.	RESULTS

FIGURE 13-2

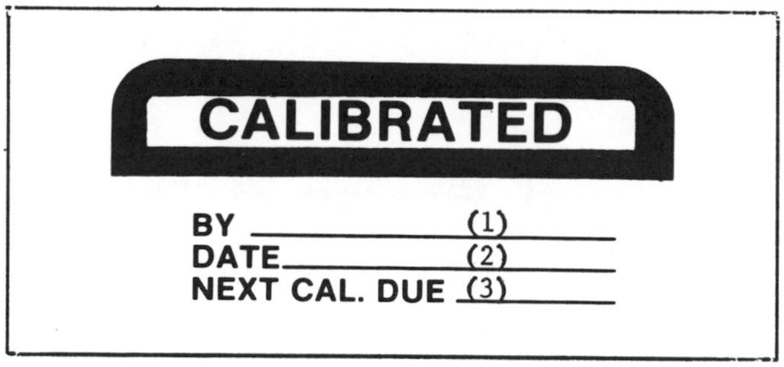

CALIBRATED

BY _____ (1) _____
DATE_____ (2) _____
NEXT CAL. DUE _(3)_ _____

FIGURE 13-3

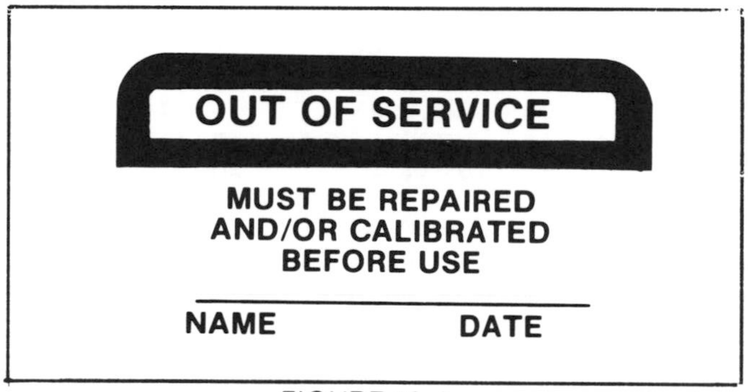

OUT OF SERVICE

MUST BE REPAIRED
AND/OR CALIBRATED
BEFORE USE
_____
NAME   DATE

FIGURE 13-4

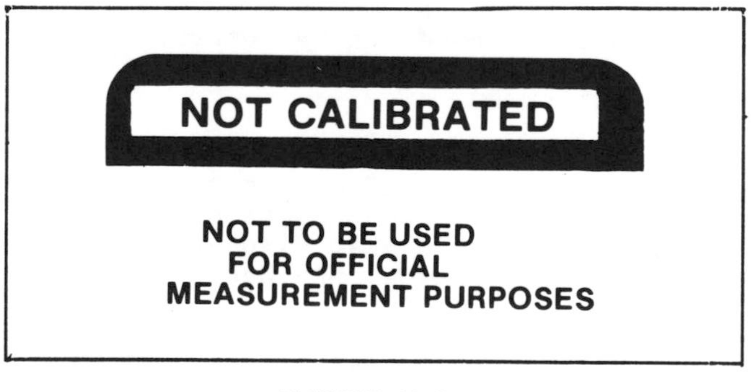

NOT CALIBRATED

NOT TO BE USED
FOR OFFICIAL
MEASUREMENT PURPOSES

FIGURE 13-5

U. S. Department of Commerce
Rogers C. B. Morton
Secretary
National Bureau of Standards
Richard W. Roberts, Director

National Bureau of Standards
Certificate of Analysis
Standard Reference Material 2676
Metals on Filter Media
(Pb, Zn, Cd, Mn)

This Standard Reference Material is intended primarily for use as an analytical standard for the determination of cadmium, lead, manganese, and zinc in the industrial atmosphere. The SRM consists of a set of membrane filters on which have been deposited the indicated quantities of salts of the particular metals.

Filter	Metal Content, μg/filter			
	Cd	Pb	Mn	Zn
A 1	0.50 ± .04	6.8 ± 1.1	1.93 ± .29	1.02 ± .06
A 2	2.48 ± .14	29. ± 2.6	10.3 ± 1.5	5.10 ± .26
A 3	10.1 ± .4	102. ± 6	20.6 ± 1.0	10.1 ± 1.1

The filters were prepared by depositing on them known and carefully reproduced volumes of solutions of pure salts, using the technique described in NBS report NBSIR 73-256.

The certified values are based upon determination of the metal content by atomic absorption spectrometry and by polarographic measurement. In these analyses, entire filters were mineralized by digestion in acid prior to measurement. The certified values are the means of those found by the two techniques while the uncertainties represent the 95 percent tolerance limits based on measurement error and variability between samples*.

The filters are identified by the numbers A1, A2, A3 printed on their edge. The metal content of the inked identification is negligible so it need not be removed. An entire filter must be used for each measurement since the metals are not uniformly distributed.

* See page 14, The Role of Standard Reference Materials in Measurement System, NBS Monograph 148, 1975. The concept of tolerance limit is also discussed in Chapter 2, Experimental Statistics, NBS Handbook 91, 1966.

In brief, if measurements were made on all the units, almost all (at least 95 percent) of these measured values would be expected to fall within the indicated tolerance limits with a confidence coefficient of 95 percent (or probability = .95).

(over)

FIGURE 15-1

The filters were prepared by R. Mavrodineanu and J. R. Baldwin. Atomic absorbtion analyses were made by them and also by T. C. Rains. E. J. Maienthal performed the polarographic analyses.

The overall direction and coordination of the technical measurements leading to certification were under the chairmanship of J. K. Taylor.

The technical and support aspects involved in certification and issuance of this Standard Reference Material was coordinated through the Office of Standard Reference Materials by W. P. Reed.

Washington, D.C. 20234   J. Paul Cali, Chief
June 30, 1975   Office of Standard Reference Materials

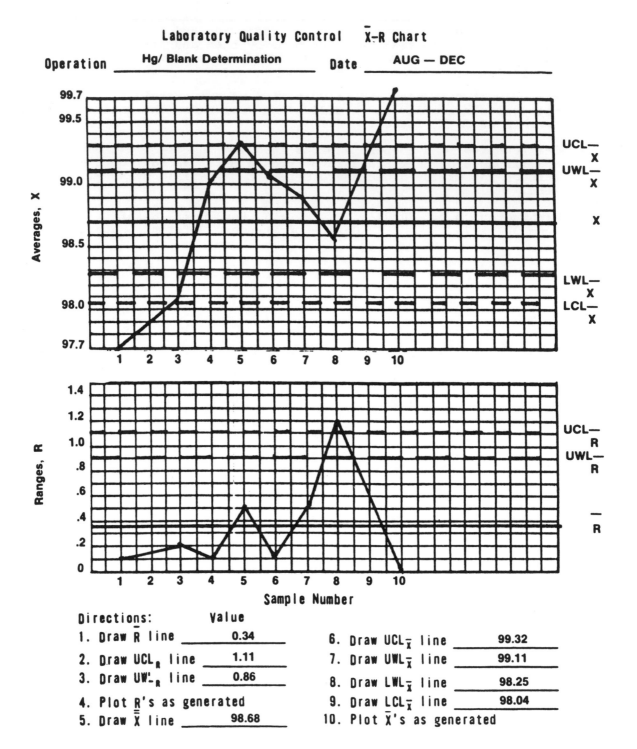

FIGURE 21-1

Laboratory Quality Control $\bar{X}$–R Chart

Operation _____   Date _____

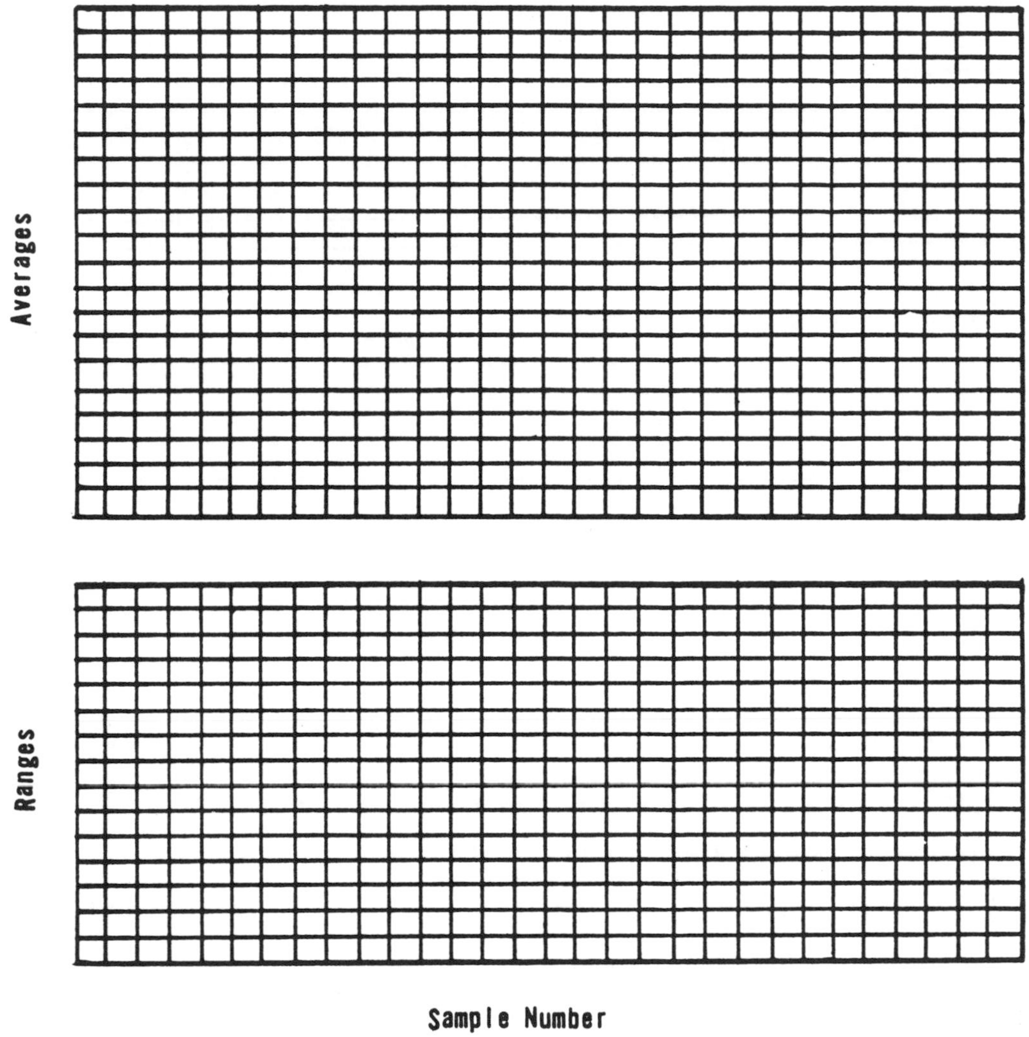

Averages

Ranges

Sample Number

Directions: Value
1. Draw $\bar{R}$ line _____   6. Draw $UCL_{\bar{X}}$ line _____

2. Draw UCL_R line _____   7. Draw $UWL_{\bar{X}}$ line _____
3. Draw UWL_R line _____   8. Draw $LWL_{\bar{X}}$ line _____

4. Plot R's as generated   9. Draw $LCL_{\bar{X}}$ line _____
5. Draw $\bar{\bar{X}}$ line _____   10. Plot $\bar{X}$'s as generated

FIGURE 21-2

NONCONFORMANCE EVENT REVIEW REPORT

SAMPLE IDENT. NO. _____   DATE _____

SAMPLE DESCRIPTION _____
_____

METHOD _____
_____

SUSPECTED NONCONFORMANCE _____
_____
_____

EXPECTED RESULT _____

ACTIONS TAKEN

1. REPORTED TO: _____

2. REPORTED BY: _____

2. EVENT LOGGED IN LABORATORY NOTEBOOK: YES( ) NO( )

3. CORRECTIVE ACTION REQUEST INITIATED: YES( ) NO( )   (ATTACH COPY)

4. NEW SAMPLES REQUESTED: YES( ) NO( )

5. CUSTOMER NOTIFIED: YES( ) NO( )

6. RETEST OR REANALYSIS NECESSARY: YES( ) NO( )
   IF YES, TO WHOM ASSIGNED _____

7. THIRD PARTY LABORATORY CONFIRMATION NECESSARY: YES( ) NO( ) IF YES, NAME OF
   LABORATORY _____

REVIEW DATE _____   _____

                                               QUALITY CONTROL COORDINATOR

FIGURE 22-1

200

CORRECTIVE ACTION REQUEST

Corrective Action Request Form No.	Suspense Date:

Originator	Date
Person responsible for replying	Organization

Nature of problem below:

P I
R D
O E
B N
L T
E I
M F
    I
    C   Cause of problem:
    A
    T
    I
    O
    N

Signature

During what phase of operation was the problem identified?

R   TO: Name
E   Address
T   By: (Date)
U
R
N   Information copies to: (Names and addresses)

FIGURE 23-1

201

Corrective Action Request Form No. _____

NOTE: Please return this form to the Originator with the corrective action planned on or before commitment date. If you determine that the action required is beyond the scope of your position, please indicate below who is responsible.

Corrective Action Planned. Please indicate effectivity point and dates.

C O R R E C T I V E   A C T I O N

_____
_____
_____
_____
_____
_____

Is this a temporary action?   Yes ( ) No ( )
Is this a permanent action?   Yes ( ) No ( )

Signature_____

Date _____

CAR
FINAL DISTRIBUTION
Original: Originator

ACCEPTANCE OF CA

Initially
Approved By:_____

Date:_____

Suspense Data
for Follow-up _____

Final
Approved By:_____

Date:_____

THIS SPACE
RESERVED FOR
FOLLOW—UP

CORRECTIVE ACTION REQUEST
MASTER LOG

Corrective Action Request Form No.					
Date Submitted					
Date Received					
Submitted By					
General Problem Description					
Suspected Cause					
Assigned To					
Answer Due					
Nature of Problem Investigation					
Cause of Problem					
Corrective Action					
Corrective Action Responsibility					
Originator Notified					
Demonstration of Effectiveness					
Final Close-Out					

FIGURE 23-2

QUALITY CONTROL
QUALITY SYSTEM AUDIT CHECK LIST

INSTRUCTIONS

Laboratory compliance with each question will be indicated by placing 0, 1, 2, 3, or NA on the line opposite each question.

0. Required but not being done.
1. Below acceptance standards.
2. Adequate with minor departures from good practices.
3. Adequate in all respects.

NA. (Not Applicable) when it is obvious that a question would be inappropriate, due to type activity, lack of facilities, special methods or other valid reasons.

At the conclusion of the survey, each category of questions should be analyzed and summarized for acceptability. Space is provided in the index for entering these results. Remarks or comments by the survey team will be placed on the blank page opposite the question. The following method shall be used in summarizing each category: Total the survey points for all applicable questions. Divide this total by the total applicable questions. Multiplied by the weighting factor of 3, this computation will result in a percent acceptability figure.

$$\frac{\text{Total points for applicable questions rated}}{\text{Number of applicable questions X 3}} = \text{Percent Acceptability}$$

Example:

For category 1, we have a total of 10 applicable questions and a weighting factor of 3 per question (10 x 3)

Therefore $\frac{28}{10 \times 3}$ 93% (Percent acceptability for category)

SURVEY BY AND DATE

_____ SURVEY RESULTS _____ OF _____ = _____ %

_____

_____

_____

NAME — TITLE — DATE

FIGURE 25-1

QUALITY SYSTEM AUDIT CHECK LIST

INDEX

CATEGORY	SUBJECT	SURVEY RESULTS		
		OF		%
1	Quality Goals and Objectives			
2	Quality Policies			
3	The Quality Organization			
4	Management of the Quality Manual			
5	Quality Planning			
6	Quality in Procurement			
7	Sample Handling, Storage and Shipping			
8	Chain-of-custody Procedures			
9	Laboratory Testing and Analysis Control			
10	Quality Documentation and Records Control			
11	Control of Measuring and Test Equipment			
12	Preventive Maintainance			
13	Reference Standards			
14	Data Validation			
15	Environmental Control			
16	Customer Complaints			
17	Subcontracting			
18	Personnel Training, Qualification & Motivation			
19	Statistical Methods			
20	Nonconformity			
21	Corrective Action			
22	Quality Cost Reporting			
23	Quality Audits			
24	Reliability			

FIGURE 25-1

CATEGORY 1

QUALITY GOALS AND OBJECTIVES

1. Are the Quality Objectives clearly stated?   _____

2. Are the Quality Objectives quantified insofar as

    possible?   ____

3. Have the stated Quality Objectives been approved

    by management?   _____

                                Total Applicable Points_____

                                Total Survey Points   _____

                                Percentage   _____

CATEGORY 2

QUALITY POLICIES

1. Are the Quality Policies clearly stated?   _____

2. Do the Quality Policies support the

    achievement of quality objectives previously

    stated?   _____

3. Are the Quality Policies approved by management? _____

                                Total Applicable Points_____

                                Total Survey Points   _____

                                Percentage   _____

CATEGORY 3

THE QUALITY ORGANIZATION

1. Laboratory management reflects a positive
   attitude toward Quality Control   _____

2. Direct responsibility for Quality Control has been
   formally established.   _____

3. The functional relationship between Quality
   Control and other laboratory components has been
   clearly delineated.   _____

4. The line of authority within the Quality Control
   organization has been clearly delineated.   _____

5. The job description for the Quality Control
   Coordinator is up-to-date and accurately describes
   the current duties, responsibilites and authority of
   that individual.   _____

6. The Quality Control Coordinator is not subordinate
   to any individual who is responsible for carrying
   out testing or analytical responsibilities.   _____

                              Total Applicable Points _____

                              Total Survey Points   _____

                              Percentage   _____

CATEGORY 4

MANAGEMENT OF THE QUALITY MANUAL

1. The responsibility for maintenance of the Quality
    Manual is clearly established.   _____

2. The issue of the manual is controlled.   _____

3. Provision is made for updating the Manual and
    issuing revisions to holders of controlled
    copies.   _____

4. Provision is made for the issue of uncontrolled
    copies as necessary.   _____

                              Total Applicable Points _____

                                Total Survey Points   _____

                                Percentage   _____

CATEGORY 5

QUALITY PLANNING

1. The responsibility for the quality planning
    function is clearly defined.   _____

2. There is objective evidence that quality
    planning is an on-going, continuous process.   _____

3. The Quality Manual accurately reflects current
    quality system activities in the laboratory.   _____

                              Total Applicable Points_____

                                Total Survey Points   _____

                                Percentage   _____

CATEGORY 6

QUALITY IN PROCUREMENT

1. Inspection is performed by Quality Control. _____

2. Inspection instructions are available to
    inspection personnel. _____

3. Inspection personnel perform inspection operations
    in accordance with current instructions and
    procedures. _____

4. There is a system to maintain inspection
    instructions current. _____

5. There is a program for Quality Control approval
    of vendors and subcontractors. _____

6. Vendors and subcontractors are required to have a
    a system for quality control. _____

7. Inspection and test results are used to evaluate
    vendor quality performance. _____

8. Purchase orders, specifications, and vendors'
    catalogs are available to receiving inspection. _____

9. Inspection gages and test equipment are adequate to
    perform required inspection. _____

10. Sampling inspection (when applicable) is performed
    to ANSI/ASQC Z 1.4-1987, ANSI/ASQC Z 1.9-1987,
    or other approved sampling plan. _____

11. Approved laboratories are used for special
    tests and analyses outside the scope of the
    laboratory's capabilities. _____

12. All reagent stock is properly labeled, dated and
    marked with shelf-life expiration when necessary. _____

13. Chemical/physical analysis, test specifications,
    or certifications are required when applicable. _____

                                    Total Applicable Points_____

                                    Total Survey Points   _____

                                    Percentage   _____

CATEGORY 7

SAMPLE HANDLING, STORAGE AND SHIPMENT

1. The Quality Control Coordinator monitors the
   physical condition of incoming test samples. _____

2. Test and analytical samples are properly
   identified. _____

3. Sample Submittal Forms are completely and
   accurately filled out. _____

                                    Total Applicable Points_____

                                    Total Survey Points   _____

                                    Percentage   _____

CATEGORY 8

CHAIN-OF-CUSTODYPROCEDURES

1. Is the Chain-of-Custody Documentation Form correctly
   and completely filled out? _____

2. Does the Chain-of-Custody Documentation Form
   establish custody of the sample at all times? _____

                                    Total Applicable Points_____

                                    Total Survey Points   _____

                                    Percentage   _____

CATEGORY 9

LABORATORY TESTING AND ANALYSIS CONTROL

1. Function checks are routinely performed by the operator. _____

2. The operator performs control checks during the analytical or testing procedures. _____

3. Tests for significance of difference are made when results indicate the need. _____

4. Control charts are used routinely for suspected or difficult method determinations. _____

5. The Quality Control Coordinator is responsible for deciding which testing or analytical operations will be charted. _____

Total Applicable Points _____

Total Survey Points _____

Percentage _____

CATEGORY 10

QUALITY DOCUMENTATION AND RECORDS CONTROL

1. The quality documents to be controlled are listed in the Quality Manual. _____

2. Document retention policies are published in in the Quality Manual or elsewhere in writing. _____

3. The Quality System is adequately documented. _____

4. The Quality Control Coordinator issues periodic reports to management which reflect quality levels, quality costs, customer complaints, etc. _____

5. There is adequate control over the issue of new

   technical documents and changes.   _____

6. Obsolete documents are retrieved from work stations

   and are not permitted to remain in the hands of

   users.   _____

                              Total Applicable Points_____

                              Total Survey Points   _____

                              Percentage   _____

CATEGORY 11

CONTROL OF MEASURING AND TEST EQUIPMENT

1. Quality Control procedures insure that instruments,

   gages, measuring devices and test equipment are

   periodically inspected and recalibrated at

   established intervals.   _____

2. Working and reference standards of required

   accuracy that are periodically calibrated to

   primary standards traceable to the NIST are

   maintained in a suitable environment.   _____

3. New or repaired test equipment and measuring

   devices are calibrated and proved satisfactory

   by Quality Control before release for use.   _____

4. There are written calibration procedures for each

   type of instrument and measuring device.   _____

5. Objective evidence includes calibration date and

   date next calibration must be performed on

   measuring instrumentation.   _____

212

6. Record system provides for automatic recall of gages, instruments and test equipment due for calibration. _____

                                                  Total Applicable Points _____

                                                  Total Survey Points _____

                                                  Percentage _____

CATEGORY 12

PREVENTIVE MAINTENANCE

1. There is a planned, scheduled program tied in with the calibration system, that ensures that all measuring and test equipment undergoes periodic preventive maintenance. _____

2. Preventive maintenance checklists are in use and on file. _____

3. An adequate spare parts inventory is on hand. _____

                                                  Total Applicable Points _____

                                                  Total Survey Points _____

                                                  Percentage _____

CATEGORY 13

REFERENCE STANDARDS

1. Laboratory policy requires the use of reference standards, when available, _____

2. The Quality Control Coordinator keeps on file a copy of NIST Publication 260 and copies of NIST Certicates of Analysis. _____

$$\text{Total Applicable Points}$$

$$\text{Total Survey Points} \quad \text{_____}$$

$$\text{Percentage} \quad \text{_____}$$

CATEGORY 14

DATA VALIDATION

1. Data validation is routinely employed to detect outliers or spurious values. _____

2. Manual data validation techniques are the only methods used to detect outliers and spurious values. _____

3. Both manual and computerized techniques are used for data validation. _____

4. Sampling plans from ANSI/ASQC Z1.4, 1981 are used when selecting samples for data validation of large quantities of data. _____

Total Applicable Points _____

Total Survey Points _____

Percentage _____

CATEGORY 15

ENVIRONMENTAL CONTROLS

1. The laboratory working environment is controlled
    to the extent prescribed by the Manual   _____

                              Total Applicable Points _____

                              Total Survey Points   _____

                              Percentage   _____

CATEGORY 16

CUSTOMER COMPLAINTS

1. One individual is charged with the responsibility
    for handling all technical complaints from
    customers, regulatory agencies, and accreditation
    organizations.   _____

2. There is a system for providing feedback to all
    individuals concerned.   _____

3. One individual is charged with the responsibility
    for initiating corrective action if this becomes
    necessary.   _____

                              Total Applicable points

                              Total Survey Points   _____

                              Percentage   _____

CATEGORY 17

SUBCONTRACTING

1. Contract laboratories are required to have a viable Quality Control System in place. _____

2. Provision is made for furnishing contract laboratories with "audit" samples to test laboratory proficiency. _____

3. When a new method is introduced, provision is made for furnishing contract laboratories with spiked samples to test the proficiency of the laboratory in performing the method. _____

Total Applicable Points _____

Total Survey Points _____

Percentage _____

CATEGORY 18

PERSONNEL TRAINING, QUALIFICATION AND MOTIVATION

1. The laboratory provides for employee indoctrination and training in Quality Control _____

2. The laboratory has a program to test the effectiveness of the training effort. _____

3. Training records are available to indicate the status of individual training. _____

4. The laboratory conducts campaigns, supported by management, to promote the achievement of superior quality output. _____

Total Applicable Points _____

Total Survey Points _____

Percentage _____

CATEGORY 19

STATISTICAL METHODS

1. The laboratory employs the following statistical techniques in the routine conduct of analytical and testing work:

    a. Tests of hypotheses   _____

    b. Treatment of outliers   _____

    c. Control Charts   _____

    d. Sampling Plans   _____

    e. Tests for significance of differences   _____

    Total Applicable Points _____

    Total Survey Points   _____

    Percentage   _____

CATEGORY 20

NONCONFORMITY

1. The Quality Control System assures that positive action is taken, when a nonconforming event occurs, to ensure that all required tasks are performed in accordance with provisions of the Quality Manual.   _____

2. Adequate records are kept of nonconformities.   _____

    Total Applicable Points   _____

    Total Survey Points   _____

    Percentage   _____

CATEGORY 21

CORRECTIVE ACTION

1. The laboratory has a method of recurrence control to prevent repetitive discrepancies.   _____

2. Corrective action requests on nonconforming events are initiated promptly.   _____

3. Records provide for follow-up on all corrective action requests.   _____

4. The responsibilty for solving the problem by taking corrective action is clearly established.   _____

5. The Suspense Date when the corrective action is expected to be completed is clearly established.   _____

6. The Corrective Action Log controls the progress of the corrective action initiated by the Corrective Action Request.   _____

Total Applicable Points   _____

Total Survey Points   _____

Percentage   _____

CATEGORY 23

QUALITY COST REPORTING

1. Quality costs are accumulated for the following
    categories:

    a.   Prevention costs   _____

    b.   Appraisal costs   _____

    c.   Internal failure costs   _____

    d.   External failure costs   _____

2. When actual costs figures cannot be obtained from
    the cost accounting component, reasonably
    accurate estimates are used.   _____

3. Quality Cost reports are made to management
    periodically.   _____

4. Quality cost figures are used in the annual
    budgeting process.   _____

        Total Applicable Points _____

        Total Survey Points   _____

        Percentage   _____

CATEGORY 24

QUALITY AUDITS

1. Quality audits are performed on and announced, scheduled basis by individuals outside the quality organization. _____

2. All areas of the laboratory contributing to the quality of work output are audited. _____

3. The results of the audit are reported to management in writing. _____

4. Pre-audit and post-audit conferences are held prior to and after each audit. _____

Total Applicable Points _____

Total Survey Points _____

Percentage _____

CATEGORY 25

RELIABILITY

1. Reliability forecasts in terms of "Mean Time to Failure" (MTTF) are routinely provided by Quality Control for remotely situated recording instruments, _____

Total Applicable Points _____

Total Survey Points _____

Percentage _____

MAINTENANCE / SERVICE REPORT

ITEM NAME	TEST NO.	TIME STARTED
MANUFACTURER	TIME	DATE
MODEL	OPERATOR	

FAILURE CONDITIONS

LOCATION	INTERVAL SINCE LAST SERVICE

GENERAL DESCRIPTION OF FAILURE AND CAUSE

ENVIRONMENTAL CONDITIONS	DIAGNOSTIC TIME SPENT

CORRECTIVE ACTION TAKEN

FAILED COMPONENTS REPLACED

FAILED COMPONENTS REPAIRED

SERVICE REQUIREMENTS

TAMPER—PROOF SEALING REQUIRED ☐   NOT REQUIRED ☐

PREVENTIVE MAINTAINANCE COMPLETED

OPERATIONAL TEST COMPLETED

SIGNATURE

FIGURE 26-1

221

	Effective Date	Subject	Section No.
	By		Page of
	Approved		Revised

FIGURE 27-1

Index